Water Resource Recovery Modelling

Water Resource Recovery Modelling

Editors

Mathieu Spérandio, Yves Comeau and Leiv Rieger

Published by

IWA Publishing
Unit 104-105, Export Building
1 Clove Crescent
London E14 2BA
Telephone: +44 (0)20 7654 5500
Fax: +44 (0)20 7654 5555
Email: publications@iwap.co.uk
Web: www.iwapublishing.com

First published 2021
© 2021 IWA Publishing

Disclaimer
The information provided and the opinions given in this publication are not necessarily those of IWA and should not be acted upon without independent consideration and professional advice. IWA and the Editors and Authors will not accept responsibility for any loss or damage suffered by any person acting or refraining from acting upon any material contained in this publication.

British Library Cataloguing in Publication Data
A CIP catalogue record for this book is available from the British Library

ISBN: 9781789062403 (paperback)
ISBN: 9781789062410 (eBook)
ISBN: 9781789062427 (ePUB)

Contents

Editorial: Water Resource Recovery Modelling

As our infrastructure is transitioning from wastewater treatment to resource recovery, so must our models evolve to address the needs this transition brings. Nutrient recovery, energy production or neutrality, biomass specialization for new conversion pathways, green-house gas mitigation and more stringent effluent limits for water reclamation are driving new model development efforts and increasingly sophisticated applications of modelling. These new needs enlarge the range of biological, physical and chemical mechanisms that we need to consider in our models. Exchanging and capitalizing on these knowledges are key challenges for modellers that will bring benefits to design, operation, teaching and research.

In this issue of *Water Science & Technology*, nine papers were selected that contribute to various aspects of the field of modelling water resource recovery facilities (WRRFs). Seven of these were presented or directly arose from the 5th Water Resource Recovery Modelling (WRRmod; previously known as the Wastewater Treatment Modelling, WWTmod) Seminar held in Lake Beauport, Canada, in March 2018.

A review of outlook and challenges of WRRF modelling is first presented (Regmi *et al.* 2019). This collective paper was produced by a concerted effort of 24 individuals from various sectors of the wastewater industry.

Plant-wide aspects of modelling are then presented in two papers. Very low sludge retention time (SRT) and high rate activated sludge processes to minimize carbon oxidation and maximize organic carbon recovery by considering colloids and flocculation mechanisms are first presented (Hauduc *et al.* 2019a).

A general plant-wide model in which the sulfur and iron cycles were added is then presented and tested (Hauduc *et al.* 2019b).

Biofilm modelling by a zero-dimensional biofilm model (0DBFM) was developed for moving bed bioreactors (MBBRs) (Plattes 2019). Detachment of biofilm and attachment of suspended matter from the bulk liquid are considered in the model.

Inhibitory substances on biological nutrient removal systems motivated the development of a simulation framework using quaternary ammonium compounds (commonly used in industrial cleaners; Conidi *et al.* 2019). Biosorption and biodegradation by nitrifiers and heterotrophic organisms were simulated.

Two papers on process control are then presented. First, online control by prediction of ammonium and nitrate using a stochastic model is presented (Stentoft *et al.* 2019). Resulting improved aeration control was shown to reduce electricity costs and improve resource recovery.

Then, ammonia-based aeration control coupled with SRT (ABAC-SRT) to control ammonia in the activated sludge process is presented (Schraa *et al.* 2019). Energy consumption reduction over 30% can be expected compared with a traditional dissolved oxygen control method.

A compartmental model (CM) was shown to provide more realistic conditions than a conventional tank-in-series (TIS) configuration for the estimation of nitrous oxide production (Bellandi *et al.* 2019). The CM improved hydrodynamic consideration of local conditions and recirculation patterns both under steady state and dynamic conditions versus the TIS approach.

Thermal hydrolysis processes (THPs) can enhance biogas production in anaerobic digestion, reduce viscosity for improved mixing and dewatering, and reduce and sterilize cake solids. A combined energy (thermal heat and calorific) and process model was developed and applied at Blue Plains advanced WRRF (Aichinger *et al.* 2019). It was shown that dynamic effects were responsible for losses in electricity production of up to 29%.

Guest Editors
Mathieu Spérandio
INSA de Toulouse, France

Yves Comeau
Polytechnique Montréal, Canada

Leiv Rieger
InCTRL Solutions Inc., Canada

REFERENCES

Aichinger, P., DeBarbadillo, C., Al-Omari, A. & Wett, B. 2019 'Hot topic' – combined energy and process modeling in thermal hydrolysis systems. *Water Science & Technology* **79** (1), 84–92.

Bellandi, G., De Mulder, C., Van Hoey, S., Rehman, U., Amerlinck, Y., Guo, L., Vanrolleghem, P. A., Weijers, S., Gori, R. & Nopens, I. 2019 Tanks in series versus compartmental model configuration: Considering

doi: 10.2166/wst.2019.059

hydrodynamics helps in parameter estimation for an N2O model. *Water Science & Technology* **79** (1), 73–83.

Conidi, D., Andalib, M., Andres, C., Bye, C., Umble, A. & Dold, P. L. 2019 Modeling quaternary ammonium compound inhibition of biological nutrient removal activated sludge. *Water Science & Technology* **79** (1), 41–50.

Hauduc, H., Al-Omari, A., Wett, B., Jimenez, J., De Clippeleir, H., Rahman, A., Wadhawan, T. & Takacs, I. 2019a Colloids, flocculation and carbon capture - A comprehensive plant-wide model. *Water Science & Technology* **79** (1), 15–25.

Hauduc, H., Wadhawan, T., Johnson, B., Bott, C., Ward, M. & Takács, I. 2019b Incorporating sulfur reactions and interactions with iron and phosphorus into a general plantwide model. *Water Science & Technology* **79** (1), 26–34.

Plattes, M. 2019 Presentation and evaluation of the zero-dimensional biofilm model 0DBFM. *Water Science & Technology* **79** (1), 35–40.

Regmi, P., Stewart, H., Amerlinck, Y., Arnell, M., Garcia, P. J., Johnson, B., Maere, T., Miletic, I., Miller, M., Rieger, L., Samstag, R., Santoro, D., Schraa, O., Snowling, S., Takacs, I., Torfs, E., van Loosdrecht, M. C. M., Vanrolleghem, P. A., Villez, K., Volcke, E. I. P., Weijers, S., Grau, P., Jimenez, J. & Rosso, D. 2019 The future of WRRF modelling – outlook and challenges. *Water Science & Technology* **79** (1), 3–14.

Schraa, O., Rieger, O., Alex, J. & Miletić, I. 2019 Ammonia-based aeration control with optimal SRT control: improved performance and lower energy consumption. *Water Science & Technology* **79** (1), 63–72.

Stentoft, P. A., Munk-Nielsen, T., Vezzaro, L., Madsen, H., Mikkelsen, P. S. & Møller, J. K. 2019 Towards model predictive control: online predictions of ammonium and nitrate removal by using a stochastic ASM. *Water Science & Technology* **79** (1), 51–62.

The future of WRRF modelling – outlook and challenges

Pusker Regmi, Heather Stewart, Youri Amerlinck, Magnus Arnell,
Pau Juan García, Bruce Johnson, Thomas Maere, Ivan Miletić, Mark Miller,
Leiv Rieger, Randal Samstag, Domenico Santoro, Oliver Schraa,
Spencer Snowling, Imre Takács, Elena Torfs, Mark C. M. van Loosdrecht,
Peter A. Vanrolleghem, Kris Villez, Eveline I. P. Volcke, Stefan Weijers,
Paloma Grau, José Jimenez and Diego Rosso

ABSTRACT

The wastewater industry is currently facing dramatic changes, shifting away from energy-intensive wastewater treatment towards low-energy, sustainable technologies capable of achieving energy positive operation and resource recovery. The latter will shift the focus of the wastewater industry to how one could manage and extract resources from the wastewater, as opposed to the conventional paradigm of treatment. Debatable questions arise: can the more complex models be calibrated, or will additional unknowns be introduced? After almost 30 years using well-known International Water Association (IWA) models, should the community move to other components, processes, or model structures like 'black box' models, computational fluid dynamics techniques, etc.? Can new data sources – e.g. on-line sensor data, chemical and molecular analyses, new analytical techniques, off-gas analysis – keep up with the increasing process complexity? Are different methods for data management, data reconciliation, and fault detection mature enough for coping with such a large amount of information? Are the available calibration techniques able to cope with such complex models? This paper describes the thoughts and opinions collected during the closing session of the 6th IWA/WEF Water Resource Recovery Modelling Seminar 2018. It presents a concerted and collective effort by individuals from many different sectors of the wastewater industry to offer past and present insights, as well as an outlook into the future of wastewater modelling.

Key words | activated sludge model, big-data, computational fluid dynamics, dynamic simulation, modelling, wastewater

Pusker Regmi[†] (corresponding author)
Mark Miller
José Jimenez
Brown and Caldwell, Walnut Creek, CA, USA
E-mail: *pregmi@brwncald.com*

Heather Stewart[†]
Bruce Johnson
Jacobs, Denver, CO, USA

Youri Amerlinck
Eveline I. P. Volcke
Ghent University, Ghent, Belgium

Magnus Arnell
Department of Biomedical Engineering (BME),
    Division of Industrial Electrical Engineering and
    Automation (IEA),
Lund University,
P.O. Box 118, SE-221 00 Lund, Sweden
and
RISE Research Institutes of Sweden,
Gjuterigatan 1D, SE-582 73 Linköping, Sweden

Pau Juan García
Atkins, Bristol, UK

Thomas Maere
Elena Torfs
Peter A. Vanrolleghem
modelEAU, Université Laval, Canada
and
CentrEau,
Québec Water Research Center, Québec City, QC,
    Canada

Ivan Miletić
Leiv Rieger
Oliver Schraa
inCTRL Solutions Inc., Dundas, ON, Canada

Randal Samstag
Bainbridge Island, WA, USA

Domenico Santoro
Trojan Technologies, Research and Development,
3020 Gore Rd, London, ON N5 V 4T7, Canada

Spencer Snowling
Hydromantis ESS, Inc.,
407 King Street West, Hamilton, ON, Canada

Imre Takács
Dynamita, Nyons, France

doi: 10.2166/wst.2018.498

Mark C. M. van Loosdrecht
Delft University of Technology, Delft, The
    Netherlands

Kris Villez
Eawag, Swiss Federal Institute of Aquatic Science
    and Technology,
8600 Dübendorf, Switzerland
and
ETH Zürich, Institute of Environmental Engineering,
8093 Zürich, Switzerland

Stefan Weijers
Waterschap de Dommel, Eindhoven, The
    Netherlands

Paloma Grau
Ceit and Tecnun (University of Navarra), San
    Sebastián, Spain

Diego Rosso
University of California, Irvine,
Civil & Environmental Engineering Dept.,
Water-Energy Nexus Center,
Irvine, CA 92697-2175, USA

†Contributed equally to this paper.

THE NEED FOR QUESTIONING THE STATUS QUO

The wastewater industry is currently facing dramatic changes, shifting away from energy-intensive wastewater treatment towards low-energy, sustainable technologies capable of achieving energy- positive operation and resource recovery. The latter will shift the focus of the wastewater industry to the extraction of resources from the wastewater, as opposed to the conventional paradigm of treatment. Thanks to the pioneering developments of the past few decades, process models were established in the wastewater industry for designing, upgrading, and optimizing wastewater treatment plants. However, due to the ever expanding and ambitious objectives of wastewater management, the scope and structure of the process models of the next generation need to be re-defined to address new challenges. The new and wider vision for water resource recovery facilities (WRRFs) includes water sanitation, protection of water sources and the environment, energy reduction and production, and resource recovery. Conventional process models must be extended with new approaches such as thermodynamic, hydraulic, or economic models, just to name a few.

During the past few years, different approaches in the wastewater modelling field have been proposed, presenting new solutions to fulfil the new requirements for WRRFs. Approaches describing physicochemical models (Batstone *et al.* 2012), energy and economic cost models (Rahman *et al.* 2016), greenhouse gas models (Mannina *et al.* 2016) or methods about how to integrate all these aspects in a plant-wide context (Solon *et al.* 2017) have been recently proposed.

Moreover, hydrodynamics and mass transport have become a crucial point for the optimum design and operation of WRRFs or novel technologies dedicated to resource recovery. In addition to simulating hydraulic phenomena occurring in WRRFs (Samstag *et al.* 2016), computational fluid dynamics (CFD) models show a great potential in physicochemical processes where gaseous, solid, and aqueous phases interact and can be a very valuable tool for, for example, the optimization of the aeration process or the recovery of valuable products from waste streams by crystallization.

However, simplicity (given by the well-known and generally-accepted International Water Association (IWA) Activated Sludge Model (ASM), and Anaerobic Digestion Model (ADM)) vs complexity (proposed in new models with greater numbers of components, processes, and parameters) is a question for which there is no clear agreement in the modelling profession (Lizarralde *et al.* 2018). How complex should the models be? Can the more complex models be calibrated (i.e. fitted to data and process observations), or will additional unknowns be introduced? After almost 30 years using well-known IWA models, should the community move to other components, processes, or model structures like black box models, CFD techniques, etc.? Can new data sources – e.g. on-line sensor data, chemical and molecular analyses, new analytical techniques, off-gas analysis – keep up with the increasing process complexity and a growing need for process understanding? Are different methods for data management, data reconciliation, and fault detection mature enough to deal with such a large amount of data? Are the available calibration techniques able to cope with such complex models?

This paper represents a concerted and collective effort by individuals from many different sectors of the wastewater industry to offer insights, as well as an outlook into the future of wastewater modelling.

THE LIMITATIONS AND USEFULNESS OF ASMS

A useful process model has the level of complexity that is required to mimic the process aspects that are of importance to the investigation at hand. The complexity of the model is the result of several things, such as the temporal and spatial scales involved in the simulation. For example, oxygen dynamics take minutes while microbial population dynamics take weeks; the micro-scale of processes inside a floc are important but are frequently neglected (Picioreanu *et al.* 2007). The most significant impact on model complexity is the number of state variables that define relevant biological, chemical, and physical conversion processes. ASMs were developed as a tool for the design and operation of biological wastewater treatment (Henze *et al.* 2000). They are an effective process evaluation tool for determining the effluent composition as well as process requirements for a facility such as aeration demand, recycle pump flow requirement and sludge production as a function of time-varying influent characteristics. The inherent flexibility of the matrix model structure of ASMs facilitates the incorporation of additional microbial or chemical processes such as deammonification (Dapena-Mora *et al.* 2004), methane oxidation (Daelman *et al.* 2014), sulfide conversion (Lu *et al.* 2012), and cellulose conversion, which were not originally included in the ASM formulations. Three decades of applied dynamic-mechanistic models have produced a set of values for model parameters (specific growth rates, decay rates, yield coefficients, etc.). Most of these parameters are not site-specific, which allows for the application of the models to situations where no existing performance data exist (e.g. modelling treatment plants that are not yet built).

Proponents of ASMs recognize the models' limitations due to knowledge gaps or the simplification of complex processes. For example, there are still some relevant processes

such as the formation of nitrous oxide (a potent greenhouse gas (GHG)) and the conversions within the biological phosphorus removal process (due to the variability of phosphate accumulating organism metabolics (Gebremariam *et al.* 2011), which are not understood clearly enough to be put into a model for use in simulations.

Two limitations of ASMs, as clearly outlined in the literature, are (1) a lack of ability to properly account for solids retention times (SRT) and (2) an inability to predict sludge settleability. The recommended useful range of the models is SRTs from 3 to about 30 days (Henze *et al.* 1987). The mechanism of bioflocculation is not well-understood and this precludes an accurate prediction of effluent suspended solids (Jimenez *et al.* 2005, 2007). This is true at any SRT but is particularly relevant for high-rate systems running at SRTs less than 3 days (Smitshuijzen *et al.* 2016). There has been a growing interest in high-rate processes with the intent of maximizing carbon capture by bioflocculation while minimizing carbon oxidation. Although sludge settleability is relatively simple to measure, this parameter cannot be easily predicted because there is still a gap in the fundamental understanding of floc formation. Settling models have been coupled with ASMs, but they lack theoretical descriptions of observed settling behaviour. Moreover, they are oriented to predict return sludge concentrations but not effluent concentrations.

A further limitation of ASMs is that only one microbial group (one state variable) is considered for a single process (e.g. nitrification). At best, extended ASMs are used in which a distinction is made between ammonium oxidizing and nitrite oxidizing bacteria (Wyffels *et al.* 2004), while a wide range of species are able to carry out nitrification, all of which are reduced to a single organism behaviour that is reflected in model parameter settings (Vannecke & Volcke 2015). Explicitly considering this variation, however, is generally not required to achieve useful predictions of the macroscopic reactor behaviour (e.g. representation of effluent quality). Furthermore, the half-saturation values used in ASMs do not truly represent intrinsic affinity constants but rather lump the effect of diffusion and spatial gradients of local environments (Arnaldos *et al.* 2015; Baeten *et al.* 2018). Substrate diffusion is largely dependent on floc properties which are a function of the local shear conditions, cohesion forces related to the exopolymeric substances characteristics, and turbulence intensity (Chu *et al.* 2003). This aspect of ASMs limits the model's ability to predict the outcomes of processes like simultaneous nitrification-denitrification as well as the conversion of micropollutants in biological wastewater treatment systems.

Despite these limitations, with appropriate calibration, ASMs can provide accurate results based on the data entered into the model. The actual conditions at a facility, however, will rarely match the exact values used in the modelling effort. Thus, it is important that informed judgment is used regarding the applicability and accuracy of the model results. Fortunately, the *relative* behaviour of modelled states obtained from ASM-based simulations is typically quite accurate even if the numerical values themselves may not be. Understanding the *qualitative* response of the process to changes in facility operation or potential designs is often sufficient. This means that the extent and rigour of a model calibration exercise will depend on the model application.

Historically, uncalibrated ASMs have proven very useful as well. Models used for understanding mechanisms and for studying influencing factors require little or no calibration. For example, modelling N_2O emissions may not be used for exact prediction but are useful in identifying N_2O formation mechanisms, which in turn are crucial for evaluating greenhouse gas mitigation strategies. Such models are used not only for research but also for teaching purposes. The use of models for knowledge transfer cannot be understated. The process model that is shared and valued by multiple stakeholders facilitates effective communication and generation of insight. In addition, ASMs are used for knowledge continuity (e.g. when key staff members enter or leave a project team). The use of ASMs formed an instruction and training manual for this field for many novice engineers. As these engineers develop their skills, they, in turn, modify and expand existing ASMs, or develop competing model structures altogether. In this way, new knowledge is shared and strengthened over time.

It is almost always the misuse of a model that results in some users deciding that the model is not useful. This misuse results primarily from three sources: (1) a lack of understanding of the model structure and under which conditions it is valid, (2) improper calibration, normally from using default wastewater fractionation parameters, or (3) believing the model results are a perfect representation of the true system behaviour. The first and second items are reasonably well-recognized by most of the model-use community; however, the third item is often ignored or forgotten, even among frequent users of simulators. It is always essential to keep in mind the original modelling objective and the modelling assumptions defining the boundaries of applicability.

THE IMPACT OF DATA ABUNDANCE ON WRRF MODELLING

The largest impact of an increased abundance of data on WRRF modelling is in the inclusion of more data and non-traditional data sources into modelling and in the combination of various modelling technologies into a tool set that is more broadly-based and at the same time more unified than it is currently. Users and developers of models will have more opportunity to make use of much more (on-line) data and of various types (e.g. images from cameras, operational log books, spectra from analysers, outputs from acoustical sensors, etc.) that are either directly or indirectly related to components of interest and to employ many modelling methods to solve engineering and operational problems that are related to WRRF operation.

Because of this, there is a blurring of lines between technologies that are traditionally seen as separate and unique and which will bring about the development of hybrid models that use both traditional and new forms and sources of data. This suggests a change in thinking about how data are used in model development, and in current ideas about whether models are strictly data-driven or are based on mathematical statements about fundamental principles of conservation of mass, charge, and energy.

Currently, there is much-heated debate regarding modelling methodology starting with terminology conventions. These conventions imply that so-called 'black box' models are those that employ modelling methods that are in some fashion not directly accessible (i.e. are in some sense opaque), or that are not easily interpreted by the model developer and user. In contrast to this, the terminology 'white box' is used to describe models that are thought to describe fundamental principles based on an in-depth understanding of the underlying processes. The latter model types are often touted as more open and readily interpretable by the user (i.e. they are deemed to be more transparent in some way).

As part of this ongoing discussion, often the term 'black box' is considered synonymous to models that are data-driven and that are developed using algorithms and other methods that do not reference fundamental mass, charge, and energy balances. In contrast, the 'white box' terminology is deemed as equivalent to thermodynamic fundamentals or first-principles physical laws. While this terminology may be useful for certain purposes, it is misleading to consider black box models as purely data-driven and white box models as purely based on first principles. For example, activated sludge models (e.g. ASM1/2d/3), while considered first principles models, include a variety of Monod-type switching functions which are mathematically tractable but are not supported with theory. Data are used to determine their kinetic parameters and these switching functions are also used to fit models that do not use biological or chemical concepts directly or not at all such as hydrolysis modelling. Similarly, settling velocity functions are not based on theoretical first principles constructs but their mathematical forms and parameters can be deduced from properly designed experiments. Thus, this labelling of models as data-driven or first-principles-based is largely irrelevant. In contrast, the interpretability of a model remains an important factor in choosing a useful model, as discussed further below.

Choosing a suitable modelling technology to capture useful information from data is largely a function of utility. George E. P. Box, a prominent data analyst and statistician famously wrote roughly 40 years ago: 'All models are wrong, some are useful' (Box 1979). Today, this statement is especially compelling given the ever-increasing amount of data that are available to engineers for model-building and testing, and the ease of use of software tools. In examining the work of Box and his colleagues, it should be obvious that all useful models are derived from data of sufficient quality. These data are collected through carefully designed and performed experiments, plant trials, and/or database queries of one kind or another, and that all model-building is inherently data-driven with data acting as the principal conduit of information (Box *et al.* 2005). This information is ultimately sequestered in the form of a model and is used for various practical or theoretical applications. The use of data in model-making is clear in models such as the ASM family, as much as it is in developing models that are based on neural networks or multivariate statistical analysis. Data are used to determine kinetic parameters for biological processes and data are also used to fit models that do not use biological or chemical concepts directly or at all.

Ultimately, models are judged on their ability to predict events or process outcomes in a given application such as closed-loop control, plant design, etc. (which is the main practical application after achieving a better understanding of a process or system) (Box *et al.* 2015). User preference, familiarity, and ease of model use in practical applications are also key differentiating factors for engineering work as much as what the mathematical form of the model may be.

Well-constructed models should be able to provide clearly interpretable results for the model developer and user no matter what the model type is. In the case of activated sludge models, this comes in the form of mass, charge, and energy conservation equations, and the relationship to the biological process that is being studied. However, a process expert is needed to analyse the results. In other model types, for example, in the application of multivariate statistics, the interpretability can be achieved using contribution functions and plots that reveal how variables are combined to provide a certain model prediction or output (Miller *et al.* 1998). If these tools are properly set up, the user can have direct diagnostics as part of the model results. An example is the application of various types of multivariate statistical methods (Spearman's rank correlation analysis, hierarchical k-means clustering and principal component analysis) to relate N_2O emission from biological nitrogen removal systems (Vasilaki *et al.* 2018). With careful examination of modelling methods and supporting diagnostic tools, it should be evident that models of any type can be interpreted and used properly.

While the successful and meaningful application of tools for model interpretation is certainly important to the model developer, they should not be a barrier to using models of any kind nor should they be used to classify modelling technology in an unnecessary way. Instead, the focus should be given to fostering approaches to combine modelling methods and various data sources as well as the development of tools to help visualize, interpret, and interact with the calculated model results, such as using principal component analysis to assess membrane bioreactor fouling (Maere *et al.* 2012). A modern process simulator is an example of visualization and interaction between data, models, and model users. The increasing effort to create more user-friendly software points to the value of software development techniques in examining and interpreting data and model outputs. However, this development requires the combination data-centred methods with process knowledge. Moreover, as data become more and more abundant, it is imperative that students are trained in advanced data methods to truly engage in multi-disciplinary approaches that remove barriers to success and bring results to WRRFs. This emphasis on education will likely pay dividends in the long run as students become employees and create demand for more education and training by identifying new and creative ways of solving problems.

Model interpretation should be the key discussion point in a multidisciplinary forum. One example of successfully combining modelling, numerical methods, and data analysis is the use of multivariate methods in biological flux balance explorations. Here, principal components methods can be used in combining linear programming and flux balance modelling to deduce the distribution of glucose and ammonia in an *Escherichia coli* system (Sarıyar *et al.* 2006). Other examples that require knowledge of multiple fields to achieve practical results can be found in image analysis where cameras provide images that are used as data in closed-loop feedback control and in chemometric analysis (Prats-Montalbán *et al.* 2011).

In an increasingly data-rich environment, it is not likely that 'black box' or 'white box' models will overtake one-another as more data are used in model development. Instead, combinations of techniques (hybrid models that combine black and white box models in parallel or in series (Lee *et al.* 2005)) that can be applied to solve real-world problems will emerge as dominant as will better methods and ways to interpret and communicate model results. Model developers and users will be more able to use and combine models of many types in their quest to solve interesting and practical problems. As part of this, the development of models will increasingly involve mathematicians, computer scientists, systems engineers, and software developers as well as chemical, environmental and civil engineers, biochemists and biologists.

THE STATE OF DATA QUALITY

The improved computational power offered by new instrumentation hardware, pre-packaged algorithms, and models promises a future in which ubiquitous use of old and new instruments and data collection systems will continue to lead to improved management and operation of water infrastructures. In this paradigm, the data that are generated must be of such quality that the information needed for manual and automated decision-making can be easily extracted and used. Data should be collected because they are direct inputs to (1) model calibration and adjustment procedures for mass and energy balance models, (2) the development of data-driven and empirical models (soft-sensors), (3) closed-loop controls, and (4) operational decision support systems that are used both on-line and off-line. These applications that consume data can, in turn, produce other data that are used in various combinations and at various frequencies. This may include, for example, fault and event detection systems that exploit correlations and relationships among measured variables to detect a faulty control system, a bad sensor, an unexpected process problem, or a flawed laboratory procedure.

To define an appropriate level of data quality, it is typical to rely on measurements such as accuracy, precision, and timeliness of the produced data. This will likely remain so in the future with a strong focus on the extraction of reliable information from available data that are taken for a clearly defined purpose (Hotelling 1947). However, the use of data-augmentation algorithms could help increase the information contained in the available data (De Mulder *et al.* 2018). Measures of data quality can only be reasonably assessed within the context of the end-use of the data. It should be clear that the notion of data quality is not limited to one sensor or device taken individually. Discussions of data quality apply to networks of measurement devices and various measurement principles. This implies that high-quality data are fit for purpose, i.e. they express the information needed by the decision-maker (or an automated control system), and that low-quality data are not capable of this. An experimental design procedure balancing data collection cost and accuracy, ensuring that measured variables lead to the identification of key variables defined by the end-user, was proposed by Le *et al.* (2018).

Quite frequently, data are deemed of low data quality when the required information is obscured in the data carrying that information. Indeed, typical wastewater process data (i.e. from automated sensors) and laboratory data are noisy, may contain inconsistent values or biases and are often not available for periods of time or at the required sampling frequency. This can make both model identification protocols and control systems ineffective, leading to poor decisions, e.g. increasing energy use unnecessarily. Although the knowledge of process experts can be used to design and implement fail-safes and other safeguards to overcome some deficiencies in sensor data, implementing such fail-safes can be cumbersome and cannot guarantee a fail-free operation. This suggests that there is an opportunity to improve the designs of fail-safes, fall-backs and perhaps entire WRRF processes in general to include data handling and data processing systems. This could improve the dynamic performance of plants from the outset starting with plant designs that incorporate data management and control system considerations explicitly (Marlin 2000).

Currently, improvements to process and laboratory data can be achieved through effective actions that are performed as part of plant maintenance activities. This is not likely to change in the future, but these activities will probably expand in scope, for both sensor and laboratory data. To ensure overall data quality (i.e. fitness to purpose), these activities should target all aspects in the chain of data acquisition, transmission, storage, and end-use. Regular maintenance such as laboratory tests using standards, sensor cleaning and calibration should be combined with reviews of database structures and settings to ensure that useful information that is encapsulated in stored data is accessible when needed.

For example, improperly set data historian or SCADA system compression and logging frequencies can break correlation patterns in data that may be useful for extracting information on process faults and upsets. If these settings are not properly chosen, data may only reflect the effect of compression settings and not information on the underlying process at all.

An interesting discussion on the negative aspects of poorly-applied data compression schemes can be found in Kourti (2003). In her paper, Kourti describes how typical algorithms that are used to compress data in historians and databases can create artificial trends in data to such a degree that measured values on variables that are not related to one another in a process are identified as very highly correlated during data analysis. The artificial correlation that is caused by compression settings, including sampling rates, can be easily changed in software to ensure that a database or a historian provides data that are fit for use. These and other system design settings (e.g. connections between sensors and control loop inputs) should be reviewed periodically to ensure that the data management system can meet its purpose in supporting control systems and other goals by providing a reliable flow of information that is taken from the data.

Regular maintenance of data management systems should be coupled with data visualization and the use of models to improve data quality. When they are part of an SOP (standard operating procedure), consolidating data for plotting and visualization can provide valuable insight into the state of data and how reasonable data values may be. This can include summaries of data that are directly related to WRRF energy and effluent permit performance such as averages and key process indicators but it should also include raw values collected and stored from sensors (i.e. soft and physical sensors) and labs (Thomann *et al.* 2002). Models can be used as part of an automated methodology to identify, correct, and replace poor or missing data. Such systems can include automated mass balance calculations or soft-sensor-based diagnostic checks for outliers and other unusual conditions. Data-driven models and fundamental mass and energy balance models can be created and geared to specifically deal

with data quality so that the information the data carry can be used effectively. The combination of these calculations will improve the performance of critical systems that rely on data as inputs, especially if they are performed in an automated manner with little or no operator involvement.

For certain deviations from acceptable quality such as outliers or spikes, there are many algorithms and data treatment and filtering methods available to avoid labour-intensive data-cleaning procedures. While these methods can be beneficial in identifying and removing some causes of poor data, challenges still exist. For example, the impact of sensor drift on sensor measurements is typically much smaller than the impact of process variability and changes in the measured environment. As a result, drift is often difficult to detect algorithmically. In addition, hardware redundancy can be of limited value in dealing with drift (and other sensor faults) as all redundant sensors that measure the same variable may exhibit the same drift as a problematic sensor.

This suggests that relying on redundancy that is based on many diverse measurements (on different but related variables) simultaneously taken in a multivariate approach may be a better option for dealing with poor or missing data (Miletic *et al.* 2004). Since building this kind of redundancy may be cost-prohibitive in some cases due to the need for multiple sensors of different kinds, on-site inspection and reference measurement checks are often the only available option for maintaining data quality. This provides an opportunity for research and software development focused on improving data availability and quality (Rieger *et al.* 2010; Villez *et al.* 2016) or extracting valuable information from low-cost or poorly maintained sensors (Wani *et al.* 2017; Thürlimann *et al.* 2018).

THE ERA OF CHEAP AND FAST CFD MODELS

In 1972 Octave Levenspiel, in a widely-used textbook on chemical process engineering said, 'If we know what is happening within the vessel, then we are able to predict the behaviour of the vessel as a reactor. Though fine in principle, the attendant complexities make it impractical to use this approach.'

In his textbook, Levenspiel went on to describe the tanks in series (TIS) and axial dispersion models that he rightly felt were the best that could be used in his earlier era. The development of CFD methods since the 1970s changed this outlook significantly. CFD is the set of numerical schemes and analyses to solve momentum and continuity equations for fluid mechanics. These numerical methods are necessary because the partial differential equations describing the fluid mechanics of process tanks typically have no analytical solution and, hence, the fluid domain is generally discretized into a grid or mesh scheme. Instead of assuming homogeneity or symmetry in multiple dimensions, as is the case for one dimensional (1D) formulations, e.g. TIS (Levenspiel 1998) and 1D-settler models (Takács *et al.* 1991; Bürger *et al.* 2011), CFD includes more detail in the dimensions (Samstag *et al.* 2016).

Using CFD, one can now compute two- or three-dimensional (2D or 3D) velocity fields and follow interactions of reactants and products throughout a tank. This information can be used to optimize tank geometry and to improve designs and operation. TIS models have provided the computational base for biokinetic models like the IWA activated sludge models (ASM – Henze *et al.* 2000) for over 30 years. Today, by using CFD confirmed by field testing, it is demonstrated that the distribution of reactants and products within reactor tanks can vary widely across commonly-used reactor types. This work shows that CFD can provide a much more accurate description of these processes than was possible in an earlier era.

The wastewater modelling community remains computationally limited today when using CFD in combination with biokinetic modelling in biological wastewater treatment. Currently, it is known that CFD can be used to help predict the effectiveness of tank mixing and biological transformation in different geometries and locate sensors so that they can optimize control and be used as a calibration for simpler TIS and other models to improve their accuracy (Karpinska & Bridgeman 2016; Samstag *et al.* 2016). However, to provide more details and realism in the CFD models, extensions are required on multiphase flows, integrating kinetics, and adding distributions using population balance models for modelling phenomena such as bubbles (aeration and gas stripping) or granular sludge and flocculation. The need for additional features to expand the possibilities of CFD is clearly a limiting factor as it increases the simulation time considerably.

Regardless of the computational burden, CFD provides vital information for the design, upgrade, optimization, and operation of WRRFs. Beyond the obvious advantages for hydraulic design and capacity assessment, it also allows highlighting the impact of concentration variations in the reactor (Gresch *et al.* 2011; Rehman *et al.* 2017) which have been shown to have a significant impact in experimental work (Amaral *et al.* 2018) and measurement

campaigns (Bellandi *et al.* 2018). Documented cases of rapid return of investments in using CFD have been demonstrated also for water treatment applications, specifically in the area of ultraviolet treatment (Santoro *et al.* 2010) where strong gradients and coupled optics, chemistry, and hydraulics dictate photoreactor performance. Furthermore, CFD can help to unravel the ambiguous, as well as arbitrary, lumping of kinetic parameters such as half-saturation indices (Arnaldos *et al.* 2015, 2018) and improve the predictability of biokinetic models under varying operational conditions.

At the start, the CFD model is initialized with (dynamic) inputs and boundary conditions. Preferably, those inputs and boundary conditions are derived from measured data. In the end, the model is confronted with reality for validation, i.e. does it live up to the expectations and observations. This reality-check is clearly based on data and is essential in order to make CFD more than just a way to produce colourful images that are not easily integrated and combined with as ASM or other models.

While the general applicability of Moore's Law (a conjecture that suggests that the speed of computers doubles every 18 months) is uncertain, it is clear that within the last 10 years, a CFD model of a million cells that would have taken weeks to complete can now be completed overnight. As computer speeds have increased, the number of cells to get a finer mesh has increased substantially, rather than to dramatically reduce computation times using a coarser mesh. If recent trends of acceptance of CFD models for WRRF process design are any indication of things to follow, CFD will become a much more widely-used technique in the analysis of biological treatment than was possibly imagined in the early days of ASM modelling.

FIT FOR USE

The widespread adoption of modelling and simulation in the wastewater industry demonstrates the benefits of ASM-type models used in the last decades (Brdjanovic *et al.* 2015). In a wastewater system project, models can be used in all project phases, e.g. WWT management options, configuration, design, commissioning and operation (Daigger 2011). Despite the many purposes of modelling, the main objective in the industry is to assist designers, utility managers, and operators in decision making. Therefore, the benefit of any model does not increase with its complexity but rather by its ability to supply an adequate basis for decisions. While mass-balancing might be sufficient at an early project stage, detailed dynamic models with high accuracy are needed for optimizing operations and controller design, e.g. for aeration (Schraa *et al.* 2017).

The balance between model complexity and ease of use depends on the set of questions being addressed. Consequently, the trade-off between complexity and ease-of-use of a model is not something that is set once (i.e. prior to the investigation) and does not remain constant. It rather evolves together with knowledge and depends on the stage where the project or scientific investigation is, and the level of understanding of the phenomena being modelled.

A generally-accepted approach in searching for the optimum among complexity, robustness, and accuracy of a model relies on the principle of building-block-based model development. This approach calls for models where a building-block also represents the unit of complexity: as a result, simpler models (with fewer building-blocks) are typically preferred to complex ones. It should be stressed that models with more building-blocks do not always yield more accurate results. This is the case if the blocks added to the model are of marginal or no benefit to explaining the underlying physical processes, or potentially even interfere with the identifiability of the parameters utilized in the overall model. As such, it is good practice to conduct a statistical analysis to confirm whether added model complexity can indeed adequately explain the desired phenomenon with the increased number of model parameters. Failing to pass such a test would imply that the additional model parameters produce a model that is over-fitted in a certain context i.e. they model only noise or that the additional parameters are not numerically tractable with a result of poor estimates that seem insignificant (Draper & Smith 1998; Box *et al.* 2005). This can, therefore, affect the usefulness of the overall model leading to the apparent paradox that the more complex model is less predictive than the simpler one.

A model, whether in its conceptual stage or translated into its mathematical form – is a tool for facilitating the deployment of the scientific method in research. Such a process, that is cyclic in nature involves steps such as hypothesizing, predicting, testing and questioning. Therefore, a good model is the one supporting the investigator to refine hypotheses, to design experiments, to sharpen data analyses and to provide insight into results interpretation. For practitioners, the recent developments in improved wastewater process models are continually evolving in research and in practice. Important applications

such as novel treatment technologies, stricter effluent requirements, GHG emissions, and other sustainability indicators require a better understanding of the processes and more powerful models for decision support. This holds both for modelling new treatment processes and operational boundaries, as well as any output variables that are of interest. With respect to treatment process modelling, conventional model formulations fall short of capturing new processes such as aerobic granular sludge, anammox, and algae-based systems (Daigger 2011). More model complexity may be needed to describe these phenomena. Although a relatively straightforward ASM1-based model was considered sufficient for design and process analysis of granular systems (Volcke *et al.* 2012), complex models including granule formation and growth were considered necessary for obtaining improved process understanding. Another important area of recent research is on plant and system-wide modelling. The expansion of the models outside the plant allows for integrated evaluation and control of the whole system (Rauch *et al.* 2002). Furthermore, integrating life cycle analysis in the models makes it possible to assess the off plant environmental impact, e.g. for changes in use and recovery of resources (Arnell *et al.* 2017).

As urbanization forces many wastewater treatment plants to operate closer to their design capacity while facing stricter effluent standards, economic and operating margins are reduced. Several of the recent research models go towards fundamental biological and chemical modelling (Ni *et al.* 2014; Kazadi Mbamba *et al.* 2016; Vaneeckhaute *et al.* 2018). Excluding empirical parameters specific for each plant but rather using fundamental constants. A large number of state variables and parameters in these models is potentially a problem. The identifiability of parameters and possibility to directly measure them are often limited. However, initial studies show that the fundamental parameters are robust and require little or no calibration, provided the proper model components are included (Kazadi Mbamba *et al.* 2016; Vaneeckhaute *et al.* 2018). Still, the data requirements for model calibration and validation of more complex models are an issue as historical data of many needed states do not exist and large measurement campaigns are costly. Increasing the number of model equations might also affect simulation times. Depending on the project, this might be an issue for the modeler. However, the ever-increasing computer speed, or even access to cluster computers, makes it manageable in many cases.

In developing such models, an important yet often neglected aspect is the model verification step (i.e. the confirmation that the mathematical formulation used to describe the conceptual model is correctly implemented). For example, an unchecked model, i.e. a model that perhaps contains a subtle mathematical error such as wrong conversion factor, etc., could lead to unintended consequences if calibration is used to reconcile predicted and observed data with such a model. Such a model error could affect parameter estimations in a way that impairs the ability of the model to predict process outcomes faithfully outside its calibration range, despite the apparent good agreement with data within the calibration region. It is stressed here that adding model complexity does not always mean a more difficult model application. Appropriately-chosen sub-models for mixing and aeration of reaction networks may facilitate better model calibration and application, even if underlying models are more complicated.

THE FUTURE

Future model development will likely put emphasis on resource recovery (water, nutrients, organics, energy) rather than wastewater treatment. The practice of design, operation and control of resource recovery technology will need models that consider stringent objectives related to water-product quality, process performance stability, and operating costs. As models lead to a better understanding of processes, this may also lead to new and innovative resource recovery solutions.

For resource recovery to thrive, it will have to be considered from a broader perspective than technical feasibility. Unit process models for resource recovery will likely be integrated within broader frameworks (e.g. automated dynamic process control, sustainability, etc.) and at various scales (e.g. sewershed, watershed) to target combined social, economic, and environmental goals. Effective cost and price models will need to be developed for the different parts of the WRRF value chain in order to provide input to economic assessments. The life-cycle analysis will help decision-makers make environmentally sound choices on the most cost-effective process design and best process operation. These decisions can only be taken if the analysis comprehensively accounts for environmental aspects such as resilience assessments, and broader environmental impact studies as well as plant-level control and optimization efforts. No matter the modelling application or scope, better experimental designs that result in improved measurement campaigns for gathering key data will be paramount.

Whatever the future may hold for model development, increased data availability in combination with improved computational capacity will continue to shape the structure of future modelling frameworks. The newly-developing synergy between first principles and data-driven models has the potential to create very powerful tools for further innovation, development and decision support. However, balancing the efforts for model development and complexity, data collection, data quality assurance, and integration of different frameworks will be challenging and will require diverse technical skills.

REFERENCES

Amaral, A., Bellandi, G., Rehman, U., Neves, R., Amerlinck, Y. & Nopens, I. 2018 Towards improved accuracy in modeling aeration efficiency through understanding bubble size distribution dynamics. *Water Research* **131**, 346–355.

Arnaldos, M., Amerlinck, Y., Rehman, U., Maere, T., Van Hoey, S., Naessens, W. & Nopens, I. 2015 From the affinity constant to the half-saturation index: understanding conventional modeling concepts in novel wastewater treatment processes. *Water Research* **70**, 458–470.

Arnaldos, M., Rehman, U., Naessens, W., Amerlinck, Y. & Nopens, I. 2018 Understanding the effects of bulk mixing on the determination of the affinity index: consequences for process operation and design. *Water Science and Technology* **77** (3), 576–588.

Arnell, M., Rahmberg, M., Oliveira, F. & Jeppsson, U. 2017 Multi-objective performance assessment of wastewater treatment plants combining plant-wide process models and life cycle assessment. *Journal of Water and Climate Change* **8** (4), 715–729.

Baeten, J. E., van Loosdrecht, M. C. M. & Volcke, E. I. P. 2018 Modelling aerobic granular sludge reactors through apparent half-saturation coefficients. *Water Research* **146**, 134–145.

Batstone, D. J., Amerlinck, Y., Ekama, G., Goel, R., Grau, P., Johnson, B., Kaya, I., Steyer, J. P., Tait, S., Takács, I., Vanrolleghem, P. A., Brouckaert, C. J. & Volcke, E. 2012 Towards a generalized physicochemical framework. *Water Science and Technology* **66** (6), 1147–1161.

Bellandi, G., Porro, J., Senesi, E., Caretti, C., Caffaz, S., Weijers, S., Nopens, I. & Gori, R. 2018 Multi-point monitoring of nitrous oxide emissions in three full-scale conventional activated sludge tanks in Europe. *Water Science and Technology* **77** (4), 880–890.

Box, G. E. P. 1979 *Robustness in the Strategy of Scientific Model Building. Robustness in Statistics*. Academic Press, New York, NY, USA.

Box, G. E. P., Hunter, J. S. & Hunter, W. G. 2005 *Statistics for Experimenters: Design, Innovation, and Discovery*. Wiley-Interscience, Hoboken, NJ, USA.

Box, G. E. P., Jenkins, G. M., Reinsel, G. C. & Ljung, G. M. 2015 *Time Series Analysis: Forecasting and Control*. John Wiley & Sons, Hoboken, NJ, USA.

Brdjanovic, D., Meijer, S. C. F., López Vázquez, C. M., Hooijmans, C. M. & van Loosdrecht, M. C. M. 2015 *Applications of Activated Sludge Models*. IWA Publishing, London, UK.

Bürger, R., Diehl, S. & Nopens, I. 2011 A consistent modelling methodology for secondary settling tanks in wastewater treatment. *Water Research* **45** (6), 2247–2260.

Chu, K. H., van Veldhuizen, H. M. & van Loosdrecht, M. C. M. 2003 Respirometric measurement of kinetic parameters: effect of activated sludge floc size. *Water Science and Technology* **48** (8), 61–68.

Daelman, M. R. J., Van Eynde, T., van Loosdrecht, M. C. M. & Volcke, E. I. P. 2014 Effect of process design and operating parameters on aerobic methane oxidation in municipal WWTPs. *Water Research* **66**, 308–319.

Daigger, G. T. 2011 A practitioner's perspective on the uses and future developments for wastewater treatment modelling. *Water Science and Technology* **63** (3), 516–526.

Dapena-Mora, A., Van Hulle, S. W., Luis Campos, J., Méndez, R., Vanrolleghem, P. A. & Jetten, M. 2004 Enrichment of Anammox biomass from municipal activated sludge: experimental and modelling results. *Journal of Chemical Technology & Biotechnology* **79** (12), 1421–1428.

De Mulder, C., Flameling, T., Weijers, S., Amerlinck, Y. & Nopens, I. 2018 An open software package for data reconciliation and gap filling in preparation of Water and Resource Recovery Facility Modeling. *Environmental Modelling & Software* **107**, 186–198.

Draper, N. R. & Smith, H. 1998 *Applied Regression Analysis*. Wiley-Interscience, Hoboken, NJ, USA.

Gebremariam, S. Y., Beutel, M. W., Christian, D. & Hess, T. F. 2011 Research advances and challenges in the microbiology of enhanced biological phosphorus removal–a critical review. *Water Environment Research* **83** (3), 195–219.

Gresch, M., Armbruster, M., Braun, D. & Gujer, W. 2011 Effects of aeration patterns on the flow field in wastewater aeration tanks. *Water Research* **45** (2), 810–818.

Henze, M., Grady, C. P. L., Gujer, W., Marais, G. v. R. & Matsuo, T. 1987 Activated Sludge Model No. 1. *IAWPRC Scientific and Technical Report No. 1.*

Henze, M., Gujer, W., Mino, T. & Van Loosdrecht, M. C. M. 2000 *Activated Sludge Models ASM1, ASM2, ASM2d and ASM3*. IWA Publishing, London, UK.

Hotelling, H. 1947 *Multivariate Quality Control-Illustrated by air Testing of Sample Bombsights*. Techniques of Statistical Analysis, McGraw-Hill, New York, NY, USA, pp. 11–184.

Jimenez, J. A., La Motta, E. J. & Parker, D. S. 2005 Kinetics of removal of particulate chemical oxygen demand in the activated-sludge process. *Water Environment Research* **77** (5), 437–446.

Jimenez, J. A., La Motta, E. J. & Parker, D. S. 2007 Effect of operational parameters on the removal of particulate chemical oxygen demand in the activated sludge process. *Water Environment Research* **79** (9), 984–900.

Karpinska, A. M. & Bridgeman, J. 2016 CFD-aided modelling of activated sludge systems – A critical review. *Water Research* **88**, 861–879.

Kazadi Mbamba, C., Flores-Alsina, X., John Batstone, D. & Tait, S. 2016 Validation of a plant-wide phosphorus modelling approach with minerals precipitation in a full-scale WWTP. *Water Research* **100**, 169–183.

Kourti, T. 2003 Abnormal situation detection, three-way data and projection methods; robust data archiving and modeling for industrial applications. *Annual Reviews in Control* **27** (2), 131–139.

Le, Q. H., Verheijen, P. J. T., van Loosdrecht, M. C. M. & Volcke, E. I. P. 2018 Experimental design for WWTP data evaluation by linear mass balances. *Water Research* **142**, 415–425.

Lee, D. S., Vanrolleghem, P. A. & Park, J. M. 2005 Parallel hybrid modeling methods for a full-scale cokes wastewater treatment plant. *Journal of Biotechnology* **115** (3), 317–328.

Levenspiel, O. 1998 *Chemical Reaction Engineering* (3rd edn). Wiley, Hoboken, NJ, USA.

Lizarralde, I., Fernández-Arévalo, T., Ayesa, E., Flores-Alsina, X., Jeppsson, U., Solon, K., Vanrolleghem, P. A., Vaneeckhaute, C., Ikumi, D., Kazadi Mbamba, C., Batstone, D. & Grau, P. 2018 From WWTP to WRRF: A new modelling framework. In: *WRRmod 2018 6th IWA/WEF Water Resource Recovery Modelling Seminar*.

Lu, H., Wu, D., Jiang, F., Ekama, G. A., van Loosdrecht, M. C. M. & Chen, G.-H. 2012 The demonstration of a novel sulfur cycle-based wastewater treatment process: sulfate reduction, autotrophic denitrification, and nitrification integrated (SANI®) biological nitrogen removal process. *Biotechnology and Bioengineering* **109** (11), 2778–2789.

Maere, T., Villez, K., Marsili-Libelli, S., Naessens, W. & Nopens, I. 2012 Membrane bioreactor fouling behaviour assessment through principal component analysis and fuzzy clustering. *Water Research* **46** (18), 6132–6142.

Mannina, G., Ekama, G., Caniani, D., Cosenza, A., Esposito, G., Gori, R., Garrido-Baserba, M., Rosso, D. & Olsson, G. 2016 Greenhouse gases from wastewater treatment – A review of modelling tools. *Science of The Total Environment* **551-552**, 254–270.

Marlin, T. E. 2000 *Process Control – Designing Process and Control Systems for Dynamic Performance* (2nd edn). McGraw-Hill, New York, NY, USA

Miletic, I., Quinn, S., Dudzic, M., Vaculik, V. & Champagne, M. 2004 An industrial perspective on implementing on-line applications of multivariate statistics. *Journal of Process Control* **14** (8), 821–836.

Miller, P., Swanson, R. E. & Heckler, C. F. 1998 Contribution plots: the missing link in multivariate quality control. *International Journal of Mathematics and Computer Science* **8** (4), 775–792.

Ni, B.-J., Peng, L., Law, Y., Guo, J. & Yuan, Z. 2014 Modeling of nitrous oxide production by autotrophic ammonia-oxidizing bacteria with multiple production pathways. *Environmental Science & Technology* **48** (7), 3916–3924.

Picioreanu, C., Kreft, J. U., Klausen, M., Haagensen, J. A. J., Tolker-Nielsen, T. & Molin, S. 2007 Microbial motility involvement in biofilm structure formation–a 3D modelling study. *Water Science and Technology* **55** (8–9), 337–443.

Prats-Montalbán, J. M., de Juan, A. & Ferrer, A. 2011 Multivariate image analysis: a review with applications. *Chemometrics and Intelligent Laboratory Systems* **107** (1), 1–23.

Rahman, S. M., Eckelman, M. J., Onnis-Hayden, A. & Gu, A. Z. 2016 Life-cycle assessment of advanced nutrient removal technologies for wastewater treatment. *Environmental Science & Technology* **50** (6), 3020–3030.

Rauch, W., Bertrand-Krajewski, J. L., Krebs, P., Mark, O., Schilling, W., Schütze, M. & Vanrolleghem, P. A. 2002 Deterministic modelling of integrated urban drainage systems. *Water Science and Technology* **45** (3), 81–94.

Rehman, U., Audenaert, W., Amerlinck, Y., Maere, T., Arnaldos, M. & Nopens, I. 2017 How well-mixed is well mixed? Hydrodynamic-biokinetic model integration in an aerated tank of a full-scale water resource recovery facility. *Water Science and Technology* **76** (8), 1950–1965.

Rieger, L., Takács, I., Villez, K., Siegrist, H., Lessard, P., Vanrolleghem, P. A. & Comeau, Y. 2010 Data reconciliation for wastewater treatment plant simulation studies-planning for high-quality data and typical sources of errors. *Water Environment Research* **82** (5), 426–433.

Samstag, R. W., Ducoste, J. J., Griborio, A., Nopens, I., Batstone, D. J., Wicks, J. D., Saunders, S., Wicklein, E. A., Kenny, G. & Laurent, J. 2016 CFD for wastewater treatment: an overview. *Water Science and Technology* **74** (3), 549–563.

Santoro, D., Raisee, M., Moghaddami, M., Ducoste, J., Sasges, M., Liberti, L. & Notarnicola, M. 2010 Modeling hydroxyl radical distribution and trialkyl phosphates oxidation in UV – H_2O_2 photoreactors using computational fluid dynamics. *Environmental Science & Technology* **44** (16), 6233–6241.

Sarıyar, B., Perk, S., Akman, U. & Hortaçsu, A. 2006 Monte Carlo sampling and principal component analysis of flux distributions yield topological and modular information on metabolic networks. *Journal of Theoretical Biology* **242** (2), 389–400.

Schraa, O., Rieger, L. & Alex, J. 2017 Development of a model for activated sludge aeration systems: linking air supply, distribution, and demand. *Water Science and Technology* **75** (3), 552–560.

Smitshuijzen, J., Pérez, J., Duin, O. & van Loosdrecht, M. C. M. 2016 A simple model to describe the performance of highly-loaded aerobic COD removal reactors. *Biochemical Engineering Journal* **112**, 94–102.

Solon, K., Flores-Alsina, X., Kazadi Mbamba, C., Ikumi, D., Volcke, E. I. P., Vaneeckhaute, C., Ekama, G., Vanrolleghem, P. A., Batstone, D. J., Gernaey, K. V. & Jeppsson, U. 2017 Plant-wide modelling of phosphorus transformations in wastewater treatment systems: impacts of control and operational strategies. *Water Research* **113**, 97–110.

Takács, I., Patry, G. G. & Nolasco, D. 1991 A dynamic model of the clarification-thickening process. *Water Research* **25** (10), 1263–1271.

Thomann, M., Rieger, L., Frommhold, S., Siegrist, H. & Gujer, W. 2002 An efficient monitoring concept with control charts for on-line sensors. *Water Science and Technology* **46** (4–5), 107–116.

Thürlimann, C. M., Dürrenmatt, D. J. & Villez, K. 2018 Soft-sensing with qualitative trend analysis for wastewater treatment plant control. *Control Engineering Practice* **70**, 121–133.

Vaneeckhaute, C., Claeys, F. H. A., Tack, F. M. G., Meers, E., Belia, E. & Vanrolleghem, P. A. 2018 Development, implementation, and validation of a generic nutrient recovery model (NRM) library. *Environmental Modelling & Software* **99**, 170–209.

Vannecke, T. P. W. & Volcke, E. I. P. 2015 Modelling microbial competition in nitrifying biofilm reactors. *Biotechnology and Bioengineering* **112** (12), 2550–2561.

Vasilaki, V., Volcke, E. I. P., Nandi, A. K., van Loosdrecht, M. C. M. & Katsou, E. 2018 Relating N_2O emissions during biological nitrogen removal with operating conditions using multivariate statistical techniques. *Water Research* **140**, 387–402.

Villez, K., Vanrolleghem, P. A. & Corominas, L. 2016 Optimal flow sensor placement on wastewater treatment plants. *Water Research* **101**, 75–83.

Volcke, E. I. P., Picioreanu, C., De Baets, B. & van Loosdrecht, M. C. M. 2012 The granule size distribution in an anammox-based granular sludge reactor affects the conversion-Implications for modeling. *Biotechnology and Bioengineering* **109** (7), 1629–1636.

Wani, O., Scheidegger, A., Carbajal, J. P., Rieckermann, J. & Blumensaat, F. 2017 Parameter estimation of hydrologic models using a likelihood function for censored and binary observations. *Water Research* **121**, 290–301.

Wyffels, S., Van Hulle, S. W. H., Boeckx, P., Volcke, E. I. P., Van Cleemput, O., Vanrolleghem, P. A. & Verstraete, W. 2004 Modeling and simulation of oxygen-limited partial nitritation in a membrane-assisted bioreactor (MBR). *Biotechnology and Bioengineering* **86** (5), 531–542.

First received 1 July 2018; accepted in revised form 30 November 2018. Available online 7 December 2018

Colloids, flocculation and carbon capture – a comprehensive plant-wide model

Hélène Hauduc, Ahmed Al-Omari, Bernhard Wett, Jose Jimenez,

Haydee De Clippeleir, Arifur Rahman, Tanush Wadhawan and Imre Takacs

ABSTRACT

The implementation of carbon capture technologies such as high-rate activated sludge (HRAS) systems are gaining interests in water resource and recovery facilities (WRRFs) to minimize carbon oxidation and maximize organic carbon recovery and methane potential through biosorption of biodegradable organics into the biomass. Existing activated sludge models were developed to describe chemical oxygen demand (COD) removal in activated sludge systems operating at long solids retention times (SRT) (i.e. 3 days or longer) and fail to simulate the biological reactions at low SRT systems. A new model is developed to describe colloidal material removal and extracellular polymeric substance (EPS) generation, flocculation, and intracellular storage with the objective of extending the range of whole plant models to very short SRT systems. In this study, the model is tested against A-stage (adsorption) pilot reactor performance data and proved to match the COD and colloids removal at low SRT. The model was also tested on longer SRT systems where effluents do not contain much residual colloids, and digestion where colloids from decay processes are present.

Key words | biosorption, contact stabilization, organic substrate, oxidation, process modelling

Hélène Hauduc (corresponding author)
Tanush Wadhawan
Imre Takacs,
Dynamita SARL,
7 LD Eoupe, Nyons,
France
E-mail: helene@dynamita.com

Ahmed Al-Omari
Haydee De Clippeleir
DC Water,
5000 Overlook Ave. SW, Washington, DC 20032,
USA

Bernhard Wett
ARA Consult GmbH,
Unterbergerstraße 1, Innsbruck,
Austria

Jose Jimenez
Brown and Caldwell,
2301 Lucien Way, Suite 250, Maitland, FL 32751,
USA

Arifur Rahman
Freese and Nichols, Inc.,
2711 N Haskell Avenue, Suite 3300, Dallas, TX 75204,
USA

LIST OF ACRONYMS

AAA	alternating activated adsorption process
AB	adsorption bio-oxidation
CS	contact-stabilization
EPS	extracellular polymeric substances
HRAS	high-rate activated sludge
WRRF	water resource and recovery facilities
AHO	carbon adsorption heterotrophic organisms [g COD/m^3]
OHO	ordinary heterotrophic organisms [g COD/m^3]
$S_{B,mono}$	readily biodegradable substrate as monomers [g COD/m^3]
$S_{B,poly}$	readily biodegradable substrate as polymers [g COD/m^3]
VFA	volatile fatty acids [g COD/m^3]
$f_{EPS,VSS}$	EPS content of the sludge [g COD/gVSS]
X_{BIO}	sum of all active biomasses [g COD/m^3]

$\eta_{FLOC,Process}$ flocculation reduction factor due to process specifics [- in the range 0–1]

INTRODUCTION

The implementation of carbon capture technologies such as the A-stage of the adsorption bio-oxidation (AB) process and contact-stabilization (CS) as a form of high-rate activated sludge (HRAS) systems are gaining interests in water resource and recovery facilities (WRRFs). The intent of these systems is to minimize carbon oxidation and maximize organic carbon recovery and biogas potential (28% to 42% (Rahman *et al.* 2019)) through biosorption of degradable organics into the biomass at very low solids retention time (SRT of 0.1–1.0 d) and hydraulic retention time (HRT of 0.5–1.0 h).

doi: 10.2166/wst.2018.454

The removal of organics (measured as COD) in the A-stage can be attributed to intracellular storage and oxidation by heterotrophic bacteria together with the bioflocculation of suspended solids and colloids. The stored, flocculated and adsorbed organic material in the sludge can be removed in the settler without having been metabolized by bacteria (Smitshuijzen *et al.* 2016). These processes are taking place in a short period of time and their efficiency is linked to SRT, extracellular polymeric substances (EPS) and dissolved oxygen (DO) levels in the reactor (Jimenez *et al.* 2015; Rahman *et al.* 2016; Rahman *et al.* 2017).

The modelling of activated sludge processes, particularly the chemical oxygen demand (COD) transformation, has significantly evolved towards fundamental principles in the past decades from simple single-substrate models to more complex multiple-substrate models involving the description of oxidation, hydrolysis and storage phenomenon (Dold *et al.* 1980; Sin *et al.* 2005). However, these models were developed to describe COD removal in systems operating at long solids retention times (SRT) (i.e. 3 days or longer) where the readily biodegradable organic substrate (S_B) can be modelled as single substrate by a single removal kinetics, and the capture of slowly biodegradable substrate (particulate, X_B; and colloidal, C_B) by flocculation is not the rate limiting process, hence, it can be ignored in the models (Haider *et al.* 2003; La Motta *et al.* 2003). However, in high-rate systems (HRAS) as those employed in the AB process where the SRT may be well below one day, these assumptions with respect to organic substrate transformations are no longer applicable. In addition, for systems that include longer SRT processes, such as membrane bioreactors, biofilm-based systems (i.e. MBBRs, trickling filters) and digesters, flocculation of colloidal particles must be described by the same model to accurately describe the fate of organics throughout the whole plant.

By critically evaluating previous models and a wide range of experimental data from several carbon capture technologies, a new plant-wide mechanistic model was developed to describe colloidal material and EPS generation, flocculation, and intracellular storage in all typical units of a WRRF.

MODEL DEVELOPMENT

Existing high-rate activated sludge systems models

Based on a literature review, few models have been developed for HRAS processes, all extending the ASM1 model.

Haider *et al.* (2003) introduced two readily biodegradable substrates, one being consumed with a higher growth rate in A-stage. However, despite the experimental results of this study showing that HRAS processes select fast growing biomass, a single heterotrophic biomass was considered. Moreover, no colloidal material capture was considered in this model.

Smitshuijzen *et al.* (2016) proposed a model based on ASM1 without adding any new state variables for COD components. A fixed fraction of particulate substrate was considered absorbed in A-stage to count for colloidal removal. Kinetics were adapted for A-stage modelling but were not valid for a whole plant model.

Nogaj *et al.* (2015) model included EPS as a state variable, two readily biodegradable substrates for a single biomass, the production of storage products and the adsorption of colloidal material. This model required assumptions on EPS production yield, kinetics of production and degradation which affect the concentration of active biomass. In addition, the variability in EPS measurements associated with the type of methods used would be an issue for model calibration.

Model description

A plant-wide model (Sumo2) by Dynamita (Dynamita 2016) considering typical biological and physico-chemical reactions of activated sludge and anaerobic digestion was used as the 'base model'. This base model was modified to include the required components and processes in accordance with the critical review of existing models and experimental data, as summarized in Figure 1. The following is a list of model modifications with a brief description:

- Readily biodegradable substrate (S_B) is split into monomers and polymers ($S_{B,mono}$ and $S_{B,poly}$) as reported in Haider *et al.* (2003) and in Nogaj *et al.* (2015). $S_{B,mono}$ and $S_{B,poly}$ fractions depend on the wastewater characteristics and has been used here as calibration parameters based on A-stage residual rbCOD.
- A new fast growing biomass, carbon adsorption heterotrophic organisms (AHO) is added to help describe processes specific to HRAS. To reproduce the lower mineralization or loss of COD content observed in A-stage processes (Jimenez *et al.* 2015), this biomass stores $S_{B,mono}$ and volatile fatty acids (S_{VFA}) in storage product (X_{STO}), while the Ordinary Heterotrophic Organisms (OHO) consume only $S_{B,poly}$ and have a lower growth rate. The storage process is sensitive to DO conditions, consequently a reduction factor for the kinetic rate is

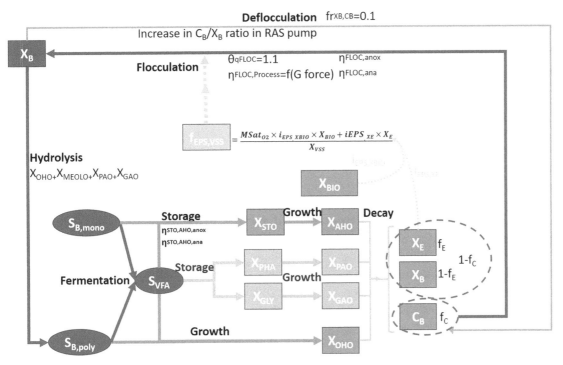

Figure 1 | Biological processes added and modified for the removal and capture of organics in the modified model.

used under anoxic and anaerobic conditions ($\eta_{STO,AHO,anox}$ and $\eta_{STO,AHO,ana}$, respectively).

- Hydrolysis is mediated only by heterotrophic organisms (OHO, MEOLO (Methylotrophs), phosphorus accumulating organisms (PAO) and glycogen accumulating organisms (GAO)) with lower growth rates to simulate the low hydrolysis observed in A-stage processes.
- Decay processes produce a fraction (f_C) of colloidal biodegradable substrate (C_B) to model the high colloidal material content observed in digesters.
- The modified model incorporates calculation of EPS content of the sludge ($f_{EPS,VSS}$) as a flocculation agent

is reduced (Nogaj *et al.* 2015). Therefore, Monod saturation functions and reduction factors for anoxic and anaerobic conditions are used in the calculation ($\eta_{FLOC,anox}$ and $\eta_{FLOC,ana}$ respectively) (Equation (1)). For lack of information and to keep the model simple as well as limiting the numbers of new parameters, it has been decided at this stage not to distinguish the impact of different biomasses in the EPS production. The role of endogenous decay products in the EPS calculation is considered in anticipation of the use of this plant-wide model in predicting granule formation, as an indicator of floc densification.

$$
\begin{aligned}
f_{EPS,VSS} = (&(Msat_{O2,KO2,EPS} \\
&+ \eta_{EPS,anox} \times Msat_{SNO2,KNO2,EPS} \times Minh_{SO2,KO2,EPS} \\
&+ \eta_{EPS,anox} \times Msat_{SNO3,KNO3,EPS} \times Minh_{SNO2,KNO2,EPS} \times Minh_{SO2,KO2,EPS} \\
&+ \eta_{EPS,ana} \times Minh_{SNO3,KNO3,EPS} \times Minh_{SNO2,KNO2,EPS} \times Minh_{SO2,KiO2,EPS}) \times i_{EPS,XBIO} \times X_{BIO} \\
&+ i_{EPS,XE} \times X_E)/X_{VSS}
\end{aligned}
\tag{1}
$$

that directly affect the flocculation kinetic rate. This calculated variable is based on the sum of calculated EPS content of biomasses ($i_{EPS,XBIO} \times X_{BIO}$) and of particulate endogenous decay products ($i_{EPS,XE} \times X_E$) related to the particulate VSS (X_{VSS}). The EPS content of biomasses (X_{BIO}) depends on electron acceptor conditions, meaning that under low DO, EPS production

where $Msat_{O2,KO2,EPS}$ and $Minh_{SO2,KO2,EPS}$ are respectively the Monod saturation and inhibition terms for O_2; $Msat_{SNO2,KNO2,EPS}$ and $Minh_{SNO2,KNO2,EPS}$ are respectively the Monod saturation and inhibition terms for NO_2; $Msat_{SNO3,KNO3,EPS}$ and $Minh_{SNO3,KNO3,EPS}$ are respectively the Monod saturation and inhibition terms for NO_3; $i_{EPS,XBIO}$ and $i_{EPS,XE}$ are respectively the EPS content of the biomass and the endogenous decay products and X_{VSS} is the particulate volatile suspended solids. The calculated extracellular polymeric substance content of the sludge ($f_{EPS,VSS}$) is an indicator of EPS production. However, the calculated values should not be compared directly to EPS measurements as results available in the literature depend on the analysis method used by different authors. In this study, the calibration has been done based on EPS measurement data obtained with cation exchange resin (CER) extraction method developed by Frølund et al. (1996); Jimenez et al. 2015).

- The kinetic rates of flocculation processes are first order reactions that are found to be well correlated with the total biomass concentration (X_{BIO}) as described by Jimenez et al. (2005), using a fraction: the calculated EPS content of the sludge ($f_{EPS,VSS}$). As flocculation is a surface reaction on flocs, the kinetic rates are considered to saturate as the amount of colloidal material becomes large in proportion to the biomass, therefore the kinetic rates include a Monod ratio saturation term for colloidal biodegradable substrate to biomasses ($MRsat_{CB,XBIO,KFLOC}$).

Furthermore, the flocculation rate is sensitive to temperature and to mixing intensity (G). Due to lack of measured data on the impact of aeration and mixing energy and equipment on deflocculation, a flocculation reduction factor ($\eta_{FLOC,Process}$ in the range 0–1) is introduced into the kinetic rate expression to account for deflocculation processes. The flocculation reduction factor varies in each reactor depending on the aeration and mixing technology. Initial values for this parameter are proposed by the authors, and can be adjusted to the case-study: $\eta_{FLOC,Process} = 0.1$ for fine bubbles, 0.25 for coarse bubble, 0.5 for anoxic mixed processes, 0.75 for anaerobic processes, 0.9 for an upflow sludge blanket and 1 for a flocculation tank. Equation (2) represents the flocculation rate calculation considered in the model.

$$r_{FLOC} = q_{FLOC,T} \times X_{BIO} \times f_{EPS,VSS} \times \eta_{FLOC,Process}$$
$$\times MRsat_{CB,XBIO,KFLOC} \qquad (2)$$

- An empirical pump process unit is implemented to simulate the deflocculation processes occurring in pumps. The factor for the increase in C_B/X_B and C_U/X_U ratio (fr_{CB_XB} and fr_{CU_XU}, respectively) can be adjusted depending on the pump technology.

The main parameters introduced in this new model are presented in Table 1 with their default values determined by the calibration study presented hereafter. The full Gujer matrix of the model is available as supplementary material (http://www.dynamita.com/public/models/Sumo2C.xlsm).

RESULTS AND DISCUSSION

The model was first calibrated to a HRAS (A-stage) pilot-scale reactor treating municipal raw wastewater operated at Laboratory of University of New Orleans (UNO) and the reactor was operated at an SRT and HRT range of 0.3–2 d and 0.5–1.1 hours, respectively (Jimenez et al. 2015). Later, the calibrated model was validated on Blue Plains Carbon management pilot reactor treating chemically enhanced primary treatment effluent (CEPT) and operated at an SRT range of 0.3–1.4 d (Rahman et al. 2016). Subsequently, the model was applied in different full-scale plant configurations such as Rochester (MN) Wastewater Treatment Plant (WWTP) A-stage and alternating activated adsorption process (AAA) Rottenburg WWTP A-stage to verify the performance of the model.

Model calibration: UNO A-stage pilot case study

The detailed description of the University of New Orleans (UNO) A-stage pilot can be found in Jimenez et al. (2015, 2005) and the configuration is presented in Figure 2. The results used for this calibration exercise were taken from Jimenez et al. (2015), in which effect the of SRT and DO on the system was investigated separately.

SRT effect

The SRT in the pilot plant was varied between 0.3 days and 2 days by continuously wasting settled mixed liquor from the return activated sludge (RAS) line. The SRT calculation neglects the biomass in the sedimentation tank. The main calibration was performed on this set of data. The EPS content of X_{BIO} and X_E ($i_{EPS,XBIO}$ and $i_{EPS,XE}$, respectively) were calibrated to fit the EPS curve as a function of SRT (Figure 3(a)); parameters for AHO and OHO growth were calibrated to fit observed soluble COD removal (Figure 3(b)) and flocculation parameters were calibrated to fit observed colloidal removal (Figure 3(c)). The parameters used in the model are listed in Table 1. All parameters have been calibrated on this data set. The model was found to be very sensitive to influent

Table 1 | Main parameters for OHO and AHO growth, EPS calculation and flocculation

Parameter	Sumo2 OHO	Sumo2C OHO	AHO
		Sumo2C	
Substrate		$S_{B,poly}$ or S_{VFA}	$S_{B,mono}$
Conditions of growth	ox, anox, ana	ox, anox, ana	ox
Rate of SB, mono storage into X_{STO} for AHOs, $q_{AHO,STO}$ (d^{-1})		–	6
Reduction factor for anoxicic storage of AHOs, $\eta_{STO,AHO,anox}$			0.25
Reduction factor for anaerobic storage of AHOs, $\eta_{STO,AHO,ana}$			0.6
Maximum specific growth rate, μ_{Max} (d^{-1})	4	2	6
Half-saturation of readily biodegradable substrate, K_S (g $COD.m^{-3}$)	5	5	0.5
Half-saturation of O_2, K_{O2} (g $O_2.m^{-3}$)	0.05	0.05	0.5
Decay rate, b (d^{-1})	0.62	0.2	0.6
Yield (g $X_{BIO}.$ g S_B^{-1})	0.67	0.67	0.6
N content of biomass (mg N/mg COD)	0.07	0.07	0.1
EPS content of X_{BIO}, $i_{EPS,XBIO}$ (g COD.g COD^{-1})		0.11	
EPS content of X_E, $i_{EPS,XE}$ (g COD.g COD^{-1})		0.01	
Half-saturation of O_2 for EPS, $K_{O2,EPS}$ (g $O_2.m^{-3}$)		0.5	
Half-saturation of NO_3 for EPS, $K_{NO3,EPS}$ (g $N.m^{-3}$)		0.1	
Half-saturation of NO_2 for EPS, $K_{NO2,EPS}$ (g $N.m^{-3}$)		0.05	
Reduction factor for anoxic flocculation, $\eta_{EPS,anox}$ (-)		0.75	
Reduction factor for anaerobic flocculation, $\eta_{EPS,ana}$ (-)		0.5	
Rate of flocculation, q_{FLOC} (d^{-1})		70	
Arrhenius coefficient for flocculation, θ_{qFLOC}		1.1	
Fraction of colloidal substrate produced in biomasses death, f_C (g C.g X_{BIO}^{-1})		0.1	

ox: aerobic conditions; anox: anoxic conditions; ana: anaerobic conditions.

fractionation (Table 2), especially at low SRT. With a single set of parameters, the model captured the effect of SRT on EPS production, soluble and colloidal COD removal, while the base model (Sumo2) could not (light green lines on Figure 3(b) and 3(c)).

Effect of dissolved oxygen concentration

The DO level varied between 0.1 and 2 mg O_2/L for three different SRTs: 0.5, 1 and 1.5 d. The experiments on the DO effect allowed the calibration of half-saturation parameters

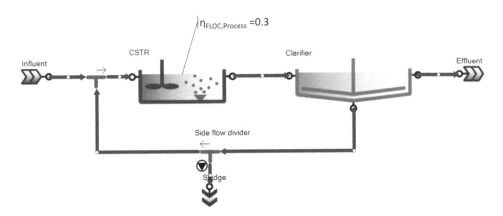

Figure 2 | University of New Orleans (UNO) A-stage pilot plant configuration and flocculation factors ($\eta_{FLOC,Process}$) applied.

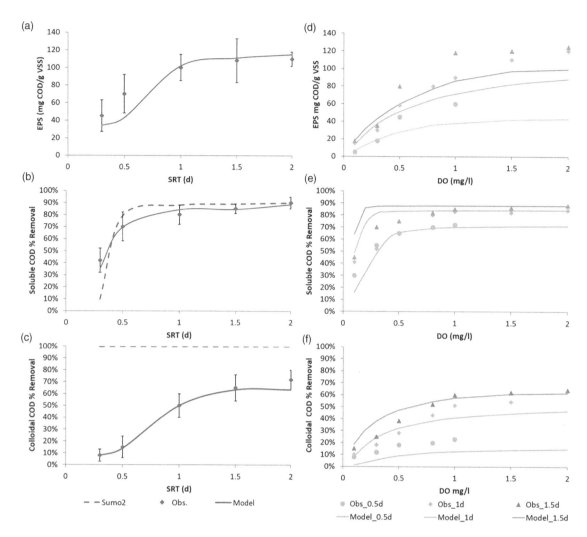

Figure 3 | Calibration results: SRT effect (a, b and c respectively for EPS, soluble COD and colloidal COD removal) and DO effect (d, e and f respectively for EPS, soluble COD and colloidal COD removal).

Table 2 | Influent characteristics of the two sets of experiments for the UNO pilot

Influent characteristics	SRT effect	DO effect
TSS [mg/L]	171	192
Total COD [mg COD/L]	395	457
Filtered COD [mg COD/L]	180	188
Flocculated filtered COD [mg COD/L]	120	119
Colloids [mg COD/L]	60	69
Fraction of heterotrophs (OHO) in total COD (TCOD) [%]	10	5
Influent carbon adsorption heterotrophs (AHO) fraction of the heterotrophs [%]	83	90
Readily biodegradable substrate as monomers (non-VFA) fraction of filtered COD [%]	75	80

for AHO growth and EPS calculation. All other parameters were kept at the values calibrated with SRT effect experiments. However, the influent characteristics and fractionation were slightly modified to better represent the quality of the influent at the time of the experiments (Table 2). It should be noted that this influent was fairly variable (Jimenez *et al.* 2015) and contains more particulate material than in the first set of the experiment. This may explain why the modelled EPS values are 33% lower than the observed ones (Figure 3(d)), as the X_{BIO}/VSS ratio of the mixed liquor suspended solids (MLSS) is reduced (Equation (1)). Nevertheless, the effect of DO on EPS production is well captured by the model. The colloidal COD removal (Figure 3(f)) is well represented at 1 d and 1.5 d SRT, while at 0.5 d SRT the flocculation efficiency is more sensitive to the calculated EPS.

The soluble COD removal (Figure 3(e)) was closely predicted for 0.5 d SRT, while at lower DO the 1 d and 1.5 d SRT modelling results gave higher removal than observed. At these SRTs, the OHO population started to grow and outcompeted the AHOs. A modification of OHOs parameters and especially an increase of the half saturation of O_2 and readily biodegradable substrate, would have resulted in better fit for this specific dataset. However, the objective was to create a valid model for the whole plant and overfitting to one dataset was avoided.

Overall, the model results were mostly sensitive to readily biodegradable substrate fractionation and influent biomass, as already noted by Nogaj et al. (2015) in their model.

Discussion on model calibration

The EPS measurement data using cation exchange resin (CER) extraction method developed by Frølund et al. (1996) is not specific to flocculent EPS. Three fractions of EPS may be defined: soluble EPS (also called soluble microbial product, SMP), loosely bound EPS (LB-EPS) and tightly bound EPS (TB-EPS), that can be fractionated with extraction methods (Li & Yang 2007; Kinyua et al. 2017). Tightly bound EPS will helps to strengthen the flocs while loosely bound EPS (LB-EPS) will allow neighbour cell attachment. EPS are mainly composed of proteins and polysaccharides, proteins being more involved in the floc structure than polysaccharides (Dignac et al. 1998). Depending on operating parameters, the ratio of proteins/polysaccharides in EPS may change, which affects the bioflocculation properties (Kinyua et al. 2017). However, Kinyua et al. (2017) show that not enough knowledge is available to understand the formation of EPS and the role of different EPS fractions in bioflocculation and floc structure. Therefore, the model presented in this study should be considered as a first step, including EPS formation and its role in bioflocculation. The model was developed to be simple for engineering use and general enough to be used to predict granulation. The model will evolve with new knowledge available in the future about EPS and bioflocculation.

Considering the EPS measurement method, the user may adapt the model calibration to their own range of measurement by adjusting the EPS content parameters (i_{EPS}) and the maximum flocculation rate (q_{FLOC}) to keep a similar order of flocculation kinetic rate value.

Based on the authors experience, the most sensitive parameters to be adjusted by the users are: (i) the flocculation reduction factor ($\eta_{FLOC,Process}$) that lumps all the hydrodynamic effects affecting the shear force on the flocs (reactors geometry, type or aerators and mixers), this acts on the residual colloids; (ii) the influent characterization and especially the fraction of heterotrophs (OHO) in total COD and the influent carbon adsorption heterotrophs (AHO) fraction of the heterotrophs, that will be crucial at very low SRT to correctly predict the MLSS and the modeled EPS; and (iii) the readily biodegradable substrate as monomers ($S_{B,mono}$) fraction of filtered COD that will drive the residual rbCOD in the A-stage.

Model validation

Blue plains carbon-management pilot

The model was validated using a different set of data from a high rate pilot at Blue Plains Advanced Wastewater Treatment Plant, Washington, DC, USA. This pilot was chosen specifically because it represented a non-conventional high rate process to test the model ability to simulate carbon redirection and capture using the same mechanisms described in the model and calibrated for a conventional process. The detailed description of the Blue Plains Carbon Management pilot reactor operated at Blue Plains Advanced WWTP can be found in Rahman et al. (2016). The pilot was unique in that it represented a scenario of having a high rate system, with two configurations such as continuously stirred tank reactor (CSTR) and high-rate contact stabilization (CS), following a chemically enhanced primary treatment (CEPT). The wastewater characteristics were different than typical A-stage applications as the colloidal fraction was low due to the upstream CEPT process. Table 3 summarizes the average wastewater characteristics of the pilot feed (CEPT effluent).

Table 3 | Average wastewater characterization of blue plains carbon management pilot feed (CEPT effluent)

Parameter	Value
Total COD (tCOD) [mgCOD/L]	157
Particulate COD (pCOD) [mgCOD/L]	80
Colloidal COD (cCOD) [mgCOD/L]	22
Flocculated and filtered COD (sCOD) [mgCOD/L]	59
Total suspended solids (TSS) [mg/L]	31
Volatile suspended solids (VSS) [mg/L]	26
Ammonia [mg N/L]	23
Total phosphorus [mg P/L]	1.5
Ortho-phosphate [mg P/L]	0.8

The pilot reactor was run at a range of SRTs and two process configurations. The model was used to simulate a CSTR configuration and high rate CS configuration. To simulate the plug-flow nature of the pilot column, a series of four CSTR were used in the model to represent one pilot column. Table 4 summarizes the scenarios where the model was used for validation. Modifications to the flocculation reduction factor and $S_{B,mono}$ fractions were conducted to closely match the pilot effluent quality in terms of COD fractions. The flocculation reduction factors were increased to 75% for contactor column and 50% for the stabilizer. The increase was justified to account for the wall effects in the pilot columns that prevent good flocculation compared to full-scale tanks. The $S_{B,mono}$ fraction was changed to 60% for scenarios with temperatures above 20 °C and 40% for scenarios with 15 °C wastewater temperature. All other parameters were the same as these used in the model calibration (Table 1). Figure 4 summarizes the effluent quality for soluble COD (sCOD) and colloidal COD (cCOD)

fractions for the pilot and the model predictions. In general, the model predicts the effluent quality for both process configurations quite well for the range of simulated SRT. The model was able to simulate the performance of colloidal capture of CS over CSTR by manipulating the flocculation coefficient, however, it also seems to be sensitive to SRT under the CS scenario where colloidal material capture under the short SRT CS scenario (0.4 d SRT) was not accurately predicted by the model.

MODEL APPLICATIONS

Rochester (MN) WWTP A-stage modelling

Rochester (MN, USA) WWTP A-stage (Figure 5) was modelled with the new model, without any parameter modification. The total SRT for the three tanks was 1.2 d. The flocculation factor in each tank has been adjusted: 0.15 for the aerated reactor, being mechanically mixed, 0.75 for the sludge blanket, which was considered as a separate tank. The RAS was returned through a screw pump, and a low increase in C_B/X_B and C_U/X_U ratio is used ($fr_{XB,CB} = 2$).

The simulation results are presented with measured data in Table 5. It shows that the new model, without any calibration of the parameters except the influent fractionation, matches the behaviour of Rochester WWTP A-stage systems.

AAA Rottenburg WWTP A-stage modelling

The AAA®-process (alternating activated adsorption; Wett *et al.* 2014) provides an alternating reactor arrangement with operation cycles involving an aeration-phase, a pre-settling phase, 50% simultaneous feeding phase into the sludge blanket and overflow discharge (at constant water level) and a phase for wasting of the settled solids directly

Table 4 | Summary of blue plains carbon management pilot reactor's different scenarios used for model validation

Parameter	Unit	CS − 0.8 d	CS − 0.4 d	CSTR − 0.8 d	CSTR − 0.2 d
Stabilizer					
Volume	L	227	227	N/A	N/A
DO	mg/L	3.6	5.4	N/A	N/A
Contactor					
Volume	L	227	227	227	227
DO	mg/L	0.36	0.5	1.5	1.3
Total SRT	Day	1.4	0.7	1.2	0.3
Aerobic SRT	Day	0.8	0.4	0.8	0.2
Wastewater temperature	Celsius	27	22	15	27

N/A-not applicable, CS-contact stabilization, CSTR-continuously stirred tank reactor.

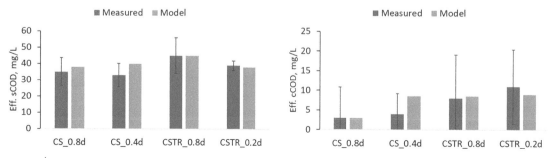

Figure 4 | Model validation against Blue Plains Carbon Management pilot performance for key parameters of effluent (Eff.) soluble COD (sCOD) and colloidal COD (cCOD) and associated standard deviations.

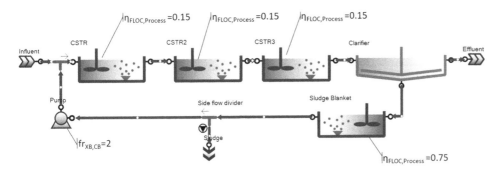

Figure 5 | Rochester (MN) WWTP A-stage configuration and flocculation factors ($\eta_{FLOC,Process}$) and deflocculation fraction ($fr_{XB,CB}$) applied for each process unit.

Table 5 | Comparison of Rochester (MN) WWTP A-stage system simulation results with measured data

	Influent	Effluent data (average of 10 d data)	Effluent model	Model error
TSS [mg/L]	66	94	94	
Filtered COD [mg COD/L]	229.1	67.2	61.3	−9%
Flocculated filtered COD [mg COD/L]	175	38.1	30.5	−20%
Colloids [mg COD/L]	54.1	29.7	31.9	6%
Total Kjeldahl nitrogen, TKN [mg N/L]	42	33.5	35.6	6%
Ammonia [mg N/L]	30	23.7	27	14%
Total phosphorus [mg P/L]	4	2.4	2.9	21%
Ortho-phosphates [mg P/L]	1.5	0.7	0.8	14%

to the adjacent thickener. The AAA-process is well suited for retrofitting existing rectangular primary clarifiers and requires an HRT of ca. 2 h. Half a year of data for winter-operation at the AAA-plant in Rottenburg (Germany), showed ca. two-thirds removal of the COD-load and one-third of the nitrogen load. From a modelling stand-point, this is an interesting case study of a system without any mechanical equipment and shear (flocculation factor of

0.9 for upflow sludge blanket) and a spatial discretization of an actually time-controlled process scheme (Figure 6). This way the model-configuration appears as a rather complex set-up of a reactive layered clarifier (upflow anoxic contactor) and a CSTR as an aerobic stabilizer, but the performance of the actually simple single-tank unit can be described very well (Table 6) without changing any stoichiometric or kinetic parameter of the model. The

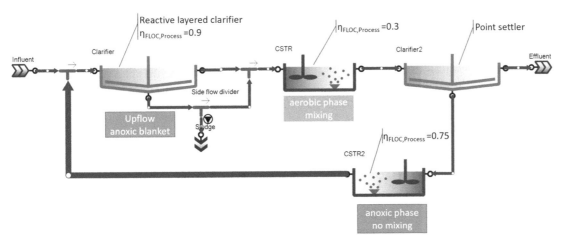

Figure 6 | AAA Rottenburg WWTP A-stage configuration and flocculation factors ($\eta_{FLOC,Process}$) applied for each process unit.

Table 6 | Comparison of AAA Rottenburg WWTP A-stage simulation results with data

	Influent	Effluent data (2 month average)	Effluent model	Model error
Total COD [mg COD/L]	356	121	122	−0.8%
Filtered COD [mg COD/L]	152	83	84.6	1.9%
Total nitrogen [mg N/L]	43.3	30.2	29.8	−1.3%
MLSS [kg/m³]		2.22	2.26	1.8%

modeled MLSS concentration is close to the observed MLSS and the nitrogen and COD removal are well described. The filtered COD contains both soluble and colloidal COD fractions. The model predicts 55% of filtered COD being colloidal, while a comparison with data is not possible due to missing flocculated filtered analysis.

CONCLUSIONS

A new model was developed to describe colloidal material removal and EPS generation, flocculation, and storage with the objective of extending the range of whole plant models to very short SRT systems, while maintaining existing and validated predictions in all other typical units of a WRRF. In this study, the model is tested against A-stage data and proved to match the COD and colloid removal at low SRT. The model was applicable for A-stage, AAA, CSTR and high-rate CS systems and was also tested on longer SRTs where effluents do not contain much residual colloids, and digestion where colloids from decay processes are present.

This new model is proposed as a new step towards a consensus model that should be able to fit the behaviour of colloidal material and biosorption in different types of configurations such as high-rate and low-rate systems including biofilm reactors (such as moving bed biofilm reactors, trickling filters) and anaerobic digesters.

SUPPLEMENTARY MATERIAL

The full Gujer matrix and parameters of the model can be downloaded at the following link: http://www.dynamita.com/public/models/Sumo2C.xlsm.

REFERENCES

Dignac, M.-F., Urbain, V., Rybacki, D., Bruchet, A., Snidaro, D. & Scribe, P. 1998 Chemical description of extracellular polymers: implication on activated sludge floc structure. *Water Sci. Technol.* **38** (8–9), 45–53. https://doi.org/10.1016/S0273-1223(98)00676-3

Dold, P., Ekama, G. & Marais, G. 1980 A general model for the activated sludge process. *Progress in Water Technology* **12** (6), 47–77.

Dynamita 2016 *Sumo User Manual*.

Frølund, B., Palmgren, R., Keiding, K. & Nielsen, P. H. 1996 Extraction of extracellular polymers from activated sludge using a cation exchange resin. *Water Res.* **30**, 1749–1758.

Haider, S., Svardal, K., Vanrolleghem, P. A. & Kroiss, H. 2003 The effect of low sludge age on wastewater fractionation (S(S), S(I)). *Water Sci. Technol.* **47**, 203–209.

Jimenez, J. A., La Motta, E. J. & Parker, D. S. 2005 Kinetics of removal of particulate chemical oxygen demand in the activated-sludge process. *Water Environ. Res. Res. Publ. Water Environ. Fed.* **77**, 437–446.

Jimenez, J., Miller, M., Bott, C., Murthy, S., De Clippeleir, H. & Wett, B. 2015 High-rate activated sludge system for carbon management – evaluation of crucial process mechanisms and design parameters. *Water Res.* **87**, 476–482.

Kinyua, M. N., Elliott, M., Wett, B., Murthy, S., Chandran, K. & Bott, C. B. 2017 The role of extracellular polymeric substances on carbon capture in a high rate activated sludge A-stage system. *Chem. Eng. J.* **322**, 428–434.

La Motta, E., Jiménez, J., Parker, D. & McManis, K. 2003 Removal of particulate COD by bioflocculation in the activated sludge process. *WIT Transactions on Ecology and the Environment* **65**.

Li, X. Y. & Yang, S. F. 2007 Influence of loosely bound extracellular polymeric substances (EPS) on the flocculation, sedimentation and dewaterability of activated sludge. *Water Res.* **41**, 1022–1030.

Nogaj, T., Randall, A., Jimenez, J., Takacs, I., Bott, C., Miller, M., Murthy, S. & Wett, B. 2015 Modeling of organic substrate transformation in the high-rate activated sludge process. *Water Sci. Technol.* **71**, 971–979.

Rahman, A., Meerburg, F. A., Ravadagundhi, S., Wett, B., Jimenez, J. A., Bott, C., Al-Omari, A., Riffat, R., Murthy, S. & De Clippeleir, H. 2016 Bioflocculation management through high-rate contact-stabilization: a promising technology to recover organic carbon from low-strength wastewater. *Water Res.* **104**, 485–496.

Rahman, A., Mosquera, M., Thomas, W., Jimenez, J. A., Bott, C., Wett, B., Al-Omari, A., Murthy, S., Riffat, R. & De Clippeleir, H. 2017 Impact of aerobic famine and feast condition on extracellular polymeric substance production in high-rate contact stabilization systems. *Chem. Eng. J.* **328**, 74–86.

Rahman, A., De Clippeleir, H., Thomas, W., Jimenez, J. A., Wett, B., Al-Omari, A., Murthy, S., Riffat, R. & Bott, C. 2019 A-stage and high-rate contact-stabilization performance comparison for carbon and nutrient redirection from high-strength municipal wastewater. *Chem. Eng. J.* **357**, 737–749.

Sin, G., Guisasola, A., De Pauw, D. J., Baeza, J. A., Carrera, J. & Vanrolleghem, P. A. 2005 A new approach for modelling simultaneous storage and growth processes for activated sludge systems under aerobic conditions. *Biotechnology and Bioengineering* **92** (5), 600–613.

Smitshuijzen, J., Pérez, J., Duin, O. & van Loosdrecht, M. C. 2016 A simple model to describe the performance of highly-loaded aerobic COD removal reactors. *Biochemical Engineering Journal* **112**, 94–102.

Wett, B., Hell, M., Andersen, M., Wellym, Fukuzaki, Y., Aichinger, P., Jimenez, J., Takacs, I., Bott, C., Murthy, S., Cao, Y. & Tao, G. 2014 Operational and structural A-stage improvements for high-rate carbon removal. In: *IWA Specialist Conference – Global Challenges: Sustainable Wastewater Treatment and Resource Recovery*, Kathmandu, Nepal.

First received 1 July 2018; accepted in revised form 16 October 2018. Available online 26 October 2018

Incorporating sulfur reactions and interactions with iron and phosphorus into a general plant-wide model

Hélène Hauduc, Tanush Wadhawan, Bruce Johnson, Charles Bott, Matthew Ward and Imre Takács

ABSTRACT

Sulfur causes many adverse effects in wastewater treatment and sewer collection systems, such as corrosion, odours, increased oxygen demand, and precipitate formation. Several of these are often controlled by chemical addition, which will impact the subsequent wastewater treatment processes. Furthermore, the iron reactions, resulting from coagulant addition for chemical P removal, interact with the sulfur cycle, particularly in the digester with precipitate formation and phosphorus release. Despite its importance, there is no integrated sulfur and iron model for whole plant process optimization/design that could be readily used in practice. After a detailed literature review of chemical and biokinetic sulfur and iron reactions, a plant-wide model is upgraded with relevant reactions to predict the sulfur cycle and iron cycle in sewer collection systems, wastewater and sludge treatment. The developed model is applied on different case studies.

Key words | anaerobic digestion, iron, sewer, sulfur, wastewater treatment, whole plant modelling

Hélène Hauduc (corresponding author)
Tanush Wadhawan
Imre Takács
Dynamita SARL,
7 LD Eoupe,
Nyons,
France
E-mail: helene@dynamita.com

Bruce Johnson
Jacobs,
Denver, CO,
USA

Charles Bott
HRSD,
Norfolk, VA,
USA

Matthew Ward
Jacobs,
Austin, TX,
USA

INTRODUCTION

Sulfur (S) comes right after the five key elements (C, O, H, N, P) by weight in activated sludge. Sulfur amounts to about 1% of dry weight of microorganisms; it is essential for life and is involved in a complex network of biological and chemical reactions. In its various forms, it can exert oxygen demand, can be used as electron acceptor or donor, can produce inert precipitates thus increasing sludge production, and can bind iron in the digester and lead to additional P release (Batstone *et al.* 2015). Furthermore, gaseous hydrogen sulfide is known to result in sulfide oxidation which results in corrosive sulfuric acid on concrete sewer surfaces exposed to air (Parker 1945). Many of the organic and inorganic forms of sulfur cause odour nuisances (WERF 2007), and corrosion in digester gas co-generation systems. Sulfur in wastewater collection systems is often handled through chemical additions (e.g. nitrate, H_2O_2, NaOCl, iron, $Mg(OH)_2$, NaOH, $Ca(OH)_2$) (Hvitved-Jacobsen *et al.* 2013), which will impact the subsequent wastewater treatment processes. Furthermore the iron reactions, resulting from addition for chemical P removal and digester gas H_2S control, interact with the

sulfur cycle, particularly in the digester with precipitate formation (Roussel & Carliell-Marquet 2016).

Despite its large importance, there was no integrated sulfur and iron model for whole plant process optimization/design that could be used in practice at the beginning of this study. However, there is a large amount of literature describing the individual biological or chemical reactions and redox transformations (Batstone *et al.* 2015). What is missing is integration with existing whole plant models and calibration to real world data and scenarios.

The objective of this work is to upgrade a whole plant model with a reliable description of sulfur and iron precipitation, volatilization, oxidation and reduction processes. The challenge in such complex systems is to keep the global model as simple as possible (i.e. limiting the number of biomasses and kinetic processes), and to develop a model able to predict the sulfur and iron cycle in sewer systems (aerobic/anoxic/anaerobic conditions with low biomass concentration), in wastewater treatment processes (aerobic/anoxic/anaerobic) and in solids treatment processes (anaerobic digesters). The sulfur and iron

doi: 10.2166/wst.2018.482

reactions implemented in the model are based on a detailed literature review.

SULFUR AND IRON BIOLOGY AND CHEMISTRY: LITERATURE REVIEW AND MODEL EXTENSION

The plant-wide model proposed in this study is a 'standard supermodel' (Grau *et al.* 2009), all processes that can occur in the collection system, in the mainstream and in the sludge line being in a single model. The main advantages of a supermodel are the absence of interfaces to connect two standard models (as would be necessary with an 'interfaces' approach using, for example, activated sludge model (ASM)-type and anaerobic digestion model (ADM)-type models) and the absence of required knowledge on the biochemical processes that might occur to adapt the model to the plant under study ('tailored supermodel'). However, the 'standard supermodel' has a higher computational cost, as all equations of the model are solved in each process unit of the configuration, which causes a minimal overload using a simulation platform with efficient algorithms.

A plant-wide model of phosphorus transformation including interlinks with sulfur and iron cycles has been proposed by Solon *et al.* (2017). This plant-wide model is based on ASM2d model for the mainstream and on ADM1 for the sludge line. Interfaces are required between the biokinetic models and the physico-chemical models. The physico-chemical model and the biochemical processes considered by Solon *et al.* (2017) are overall similar to those proposed in this study for a conventional water resource and recovery plant. However, some iron and sulfur reduction processes are not implemented in this extended ASM2d model, which could limit the applicability of the model in the case of low oxidation/reduction potential (ORP) zones (e.g. sections of collection systems, and sidestream enhanced biological phosphorus removal).

Whole plant model background

The base model is a plant-wide model (Dynamita 2016) considering typical biological and physio-chemical reactions of activated sludge and anaerobic digestion. It considers different organisms groups: (i) heterotrophs (biochemical oxygen demand (BOD) removal and denitrification in one or two steps); (ii) methanol utilizers (BOD removal under anoxic conditions); (iii) nitrifiers (one- or two-step nitrification); (iv) anammox organisms; (v) phosphate accumulating organisms, considering their behaviour under low ORP conditions and competition with glycogen accumulating organisms (Varga *et al.* 2018); and (vi) acidoclastic (AMETO) and hydrogenotrophic methanogens (HMETO).

The physio-chemical model considers: (i) chemical phosphorus removal (iron dosing, based on surface complexation model concepts (Hauduc *et al.* 2015)); (ii) precipitation reactions (forming amorphous calcium phosphate, calcium carbonate, struvite, vivianite); (iii) chemical equilibrium (ionic speciation) for pH calculation; and (iv) gas transfer.

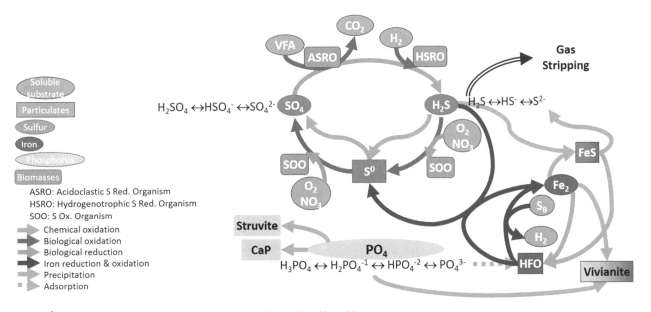

Figure 1 | Phosphorus, sulfur and iron cycles interaction implemented in the plant-wide model.

Biology and chemistry were extended with required components, species and reactions in accordance with the following literature review. Figure 1 synthesizes the sulfur cycle implemented in Sumo© models (Dynamita 2018) and the interactions with phosphorus and iron cycles.

General sulfur and iron model

The sulfur model included in Sumo includes three oxidation states of sulfur: sulfate (SO_4^{-2}) as S_{SO4}, elemental sulfur (S^0) as X_S, and sulfide (S^-) as S_{H2S}.

Considering iron, two oxidation states are included in the model. Hydrous ferric oxides (HFO) species are already included in the base model for the chemical phosphorus treatment. These state variables ($X_{HFO,H}$, $X_{HFO,L}$, $X_{HFO,old}$, $X_{HFO,H,P}$, $X_{HFO,L,P}$, $X_{HFO,H,P,old}$ and $X_{HFO,L,P,old}$ depend on the floc size and P-bound status) are considered to be the only ferric (Fe^{3+}) species in the model, as ferric iron is only minimally soluble in water (Hauduc et al. 2015). X_{HFO} is a calculated variable being the sum of the seven HFO state variables. For the ferrous iron (Fe^{2+}), a new state variable is included in the model as S_{Fe2}, and ferrous oxides are not considered.

These states are considered to interact with other wastewater components as described below.

FeS precipitation and iron interaction

Reduction of Fe^{3+}, sulfide as electron donor

A chemical reduction of Fe^{3+} by sulfide occurs under reducing conditions. In this reaction, sulfide is oxidized into colloidal elemental sulfur which precipitates (Nielsen et al. 2005; Firer et al. 2008):

$$2Fe^{3+} + HS^- \rightarrow 2Fe^{2+} + S^0 + H^+$$

Implementation in the whole plant model. Elemental sulfur has been added as a particulate state variable (X_S), as elemental sulfur has a low solubility and flocculates easily. The hydrous ferric oxides (X_{HFO}) are reduced by H_2S in a single process into ferrous iron (S_{Fe2}) and elemental sulfur (X_S) with adequate stoichiometric coefficient to balance the redox reaction and a first order rate with respect to the X_{HFO} concentration.

Reduction of Fe^{3+}, organic matter as electron donor

HFO are reduced in digesters into soluble Fe^{2+} which precipitates into iron sulfide (FeS), and release bound phosphates (Ge et al. 2013), which can further precipitate into vivianite ($Fe_3(PO_4)_2,8H_2O$) (Cheng et al. 2015). This biological process is performed by Fe^{3+}-reducing bacteria, using organic matter as electron donors (Lovley & Phillips 1988).

Implementation in the whole plant model. To keep the model simple, the small amount of iron-reducing biomass production is not introduced in the model. The soluble biodegradable substrate (S_B) and volatile fatty acids (S_{VFA}) are considered as electron donor, and a first order kinetic rate expression with respect to the X_{HFO} concentration is used.

FeS precipitation

Fe^{2+} precipitates with sulfide into FeS (Nielsen et al. 2005; Firer et al. 2008):

$$Fe^{2+} + HS^- \rightarrow FeS + H^+$$

Implementation in the whole plant model. The acid-base reactions of the sulfate and sulfide species are added in the pH model for speciation (equilibrium model). The precipitation is modelled following the Koutsoukos et al. (1980) kinetic expression with a solubility product $K_{sp,FeS} = 3.7*10^{-19}$ (Nielsen et al. 2005).

Oxidation of Fe^{2+}

According to Gutierrez et al. (2010), the precipitated FeS is re-oxidized into ferric oxides and sulfate in an aerobic zone.

Implementation in the whole plant model. Both oxidation of ferrous iron (S_{Fe2}) and precipitated iron sulfide (X_{FeS}) are considered in the model with oxygen as electron acceptor with adequate stoichiometric coefficients to balance the redox reaction and a first order rate with respect to the S_{Fe2} and X_{FeS} concentrations respectively.

Reduction of sulfate

The biological sulfate reduction is the main process step in sulfur biotreatment, often combined with a chemical step or a metal precipitation step (Hao et al. 2014). The biological sulfate reduction is performed by sulfate-reducing organisms (SRO), which can use either hydrogen or organic compounds as electron donor. These bacteria are directly in competition with hydrogenotrophic and acetoclastic methanogens respectively in anaerobic bioprocesses (Kalyuzhnyi & Fedorovich 1998; Chou et al. 2008; Hao et al. 2014) and in sewer sediments (Liu *et al.* 2016). Models for sewer systems usually neglect the biomass growth whereas models for anaerobic digestion always consider it. These models consider different kinds of substrates. Knobel & Lewis (2002) consider five substrates, and Liu *et al.* (2015) and Fedorovich *et al.* (2003) consider four substrates, whereas Batstone (2006) suggests considering only hydrogenotrophic sulfate reducer bacteria if sulfur to chemical oxygen demand (COD) ratio is below 0.1 mg S/mg COD. The best compromise seems to be the model from Barrera *et al.* (2015) and Poinapen & Ekama (2010) who consider H_2, acetate and propionate as substrates. The WATS (wastewater aerobic–anaerobic transformations in sewers) model for sewer processes (Hvitved-Jacobsen *et al.* 2013) considers only soluble substrate for sulfate reduction biological processes.

Implementation in the whole plant model. Considering the actual structure of the extended version of Sumo model, S_{VFA} and S_{H2} have been chosen as substrate for SRO, resulting in competition with the AMETO and HMETO, which would be similar to what is suggested by Barrera *et al.* (2015) and in accordance with Kalyuzhnyi & Fedorovich (1998). Similarly to the methanogenesis implementation, two biomasses are introduced: acidoclastic sulfate-reducing organisms and hydrogenotrophic sulfate-reducing organisms. This leads to four additional processes to consider growth and decay of both biomasses. Stoichiometric and kinetic values from Barrera *et al.* (2015) are used. The produced sulfide is inhibitory (Utgikar *et al.* 2002). It has been considered in the kinetic rate expression through Haldane functions when sulfide is a reactant of the process, otherwise through Monod limitation function term.

Oxidation of sulfide

Biological oxidation

The biological oxidation of sulfide into sulfate is performed through intermediate species. The oxidation may use either oxygen, nitrite or nitrate as electron acceptor. In the literature, the biological oxidation of sulfide is mainly modelled in one or two steps, S^0 being the intermediate. The oxidation of elemental sulfur to sulfate is the limiting step (Buisman *et al.* 1991; Tichy *et al.* 1998; Jiang *et al.* 2009). According to several authors, when sulfide is oxidized in a digester at limiting oxygen levels, it is converted to elemental sulfur which precipitates, making it less available for further biological reduction (Jenicek *et al.* 2008; Díaz & Fdz-Polanco 2012).

Implementation in the whole plant model. A sulfur oxidizing organism (X_{SOO}) has been introduced in the model with four oxidation processes to consider the two steps of sulfide oxidation and two possible oxidants (O_2 and NO_3). The parameter values from Mannucci *et al.* (2012) are used as first estimation.

Chemical oxidation

At high SOO activity, chemical oxidation is negligible (Luther *et al.* 2011) but must be considered in the case of sewer processes with lower biomass concentration, as the oxygen consumption for sulfide oxidation counts significantly in the oxygen uptake rate (OUR) (Nielsen *et al.* 2003). The literature reports kinetic laws with different orders and a wide range of oxidation rate parameters; however, the rate of the two steps of oxidation are not determined independently (Buisman *et al.* 1990; Nielsen *et al.* 2003; Luther *et al.* 2011; Hvitved-Jacobsen *et al.* 2013; Klok *et al.* 2013).

Implementation in the whole plant model. Two processes for oxidation of S_{H2S} by oxygen in two steps ($S_{H2S} \rightarrow X_S \rightarrow S_{SO4}$) is added. All the oxidation intermediates are considered through the elemental sulfur state variable (X_S), whereas the second oxidation step ($X_S \rightarrow S_{SO4}$) is much slower (Nielsen *et al.* 2003). To simplify the model, first order reactions with respect to sulfide and to elemental sulfur have been implemented for both steps of the oxidation process.

The full Gujer matrix of the model is available as supplementary material (http://www.dynamita.com/public/models/Sumo2S.xlsm).

RESULTS

The whole plant model behaviour is evaluated for collection system, mainstream and sludge line. All the results presented are obtained using a single set of parameters.

Sewer pipe example

A force main sewer pipe with no gas phase has been modelled by 10 segments of 1.2 h hydraulic retention time (HRT) each, so a total HRT of 12 h over the pipe. The sewer biofilm is not modelled here as it will be part of a further process unit model development. Only reactions occurring in the bulk are considered.

The simulation without any dosage shows a slight reduction of sulfate along the sewer pipe (dotted lines on

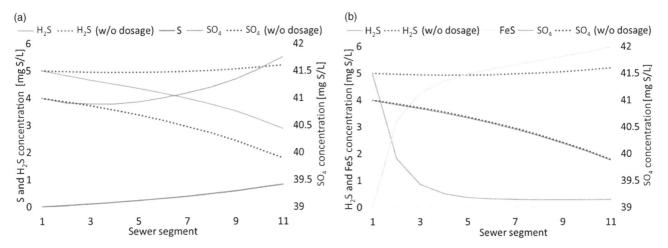

Figure 2 | Sulfur component concentrations in the sewer pipe with (a) nitrate dosage (solid lines) and (b) iron dosage (solid lines) and without dosage (dotted lines).

Figure 2) due to sulfate-reducing bacteria activity, in the absence of electron acceptors. With nitrate dosage, sulfide is oxidized into elemental sulfur (X_S) that precipitates, then into sulfate (Figure 2(a)). The accumulation of elemental sulfur confirms that in the model the oxidation of elemental sulfur to sulfate is the limiting step as stated by many authors (Buisman *et al.* 1991; Tichy *et al.* 1998; Jiang *et al.* 2009). With Fe^{2+} dosage, Fe^{2+} strongly binds sulfide into FeS, which precipitates (Figure 2(b)).

The produced compounds in each case (X_S and X_{FeS} respectively), will be either captured with primary treatments or oxidized in aerated processes of the downstream wastewater treatment plant.

Mainstream modelling

Concerning the wastewater treatment, an idealized UCT (University of Cape Town) plant was simulated. Results show sulfide oxidation into sulfate in anoxic and aerobic tanks (Figure 3) with nitrate and oxygen respectively.

The impact of sulfur on the oxygen demand has been investigated on this configuration. A first run with the base

model that does not include the sulfur implementation shows an OUR of 34.7 mg $O_2/(L \cdot h)$ for a dissolved oxygen controlled at 2 mg/L in the aerobic tank. Then, influent total sulfur concentrations from 10 mg S/L to 100 mg S/L with a sulfide fraction of 5% (0–5 mg S-H_2S/L), 10% (0–10 mg S-H_2S/L) and 20% (0–20 mg S-H_2S/L) are simulated with the new model. The influent sulfide fraction depends on the length and characteristics of the sewer network.

Figure 4 shows the impact of the influent sulfur concentration and sulfide fraction on OUR. The shape of the curve is driven by the biological oxidation of elemental sulfur to sulfate by SOO biomass, which is known to be the limiting oxidation step (Mannucci *et al.* 2012). From this result, an influent concentration of 20 mg S/L causes an underestimation of OUR of around 5% for an influent sulfide fraction from 5% to 20%. For an influent concentration of 100 mg S/L, the underestimation of OUR goes from 7% (for an influent sulfide fraction of 5%) to 17% (for an influent sulfide fraction of 20%).

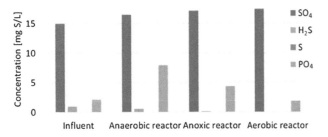

Figure 3 | Sulfur forms through the wastewater treatment (influent total sulfur of 10 mg S/L).

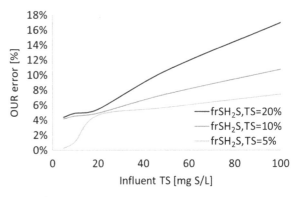

Figure 4 | Impact of influent total sulfur (TS) on OUR in the aeration tank with an influent sulfide fraction of 5%, 10% and 20% of total sulfur.

Digester modelling

A UCT plant with primary and waste sludge digestion is simulated to evaluate the impact of sulfur on the digester. A plant influent total sulfur of 20 mg/L and a S:COD ratio of 0.004 g S/g COD at the digester influent leads to a hydrogen sulfide concentration of 5,000 ppm in the biogas, which is similar to what is modelled by Flores-Alsina *et al.* (2016) for this range of S:COD ratio. The sulfate reduction and hydrogen sulfide production in the biogas are due to the biological activity of the sulfate-reducing bacteria, which can either use VFA (S_{VFA}) or dissolved hydrogen (S_{H2}) as electron acceptor in the model. They are in direct competition with methanogens, reducing the methane production from 1,289 m^3/d (model without sulfur) to 1,283 m^3/d, with respectively a methane content in the biogas of 66.1% and 64.9%.

Iron is widely used in water resource recovery facilities (WRRFs), either for phosphorus chemical treatment or to limit H_2S formation in the digester. The unintended consequence on digestion is the precipitation of iron into vivianite with phosphorus and into FeS or pyrite with sulfide. For a plant influent of 20 mg S/L, iron doses from 0 to 30 g/kg of digester dry solids are simulated to compare with Roussel & Carliell-Marquet (2016) work. Figure 5 shows the results, with first the precipitation of FeS, then of vivianite, with very close values to those of Roussel & Carliell-Marquet (2016). Figure 5 furthermore shows the decrease of hydrogen sulfide in the biogas as FeS precipitates.

The Lander Street WRRF (Boise, Idaho, USA) was experiencing problems with high hydrogen sulfide content of the biogas and struvite precipitations, resulting in increased maintenance requirements. A study was performed to evaluate the required doses of iron to solve these two problems, by precipitating sulfide and

phosphorus. A sampling programme of 45 days around the primary digester provided good knowledge of its operation performance (Table 1). The primary digester is simulated with equivalent total solids and elemental loads of nitrogen, phosphorus, sulfur, calcium and magnesium. The model results are similar to the measured data, especially concerning gas production, and hydrogen sulfide and phosphate concentrations.

For this study, doses of 25 g FeCl$_3$/kg VSS (volatile suspended solids) and of 100 g FeCl$_3$/kg VSS were estimated by experts to control respectively hydrogen sulfide and struvite precipitation. These doses were applied in the simulation and results are shown in Table 1. With dosage for H_2S control, the model correctly predicts the reduction of hydrogen sulfide by more than 10 times (from 1,926 ppm to 141 ppm)

Table 1 | Data and simulation results of Lander Street primary digester, and two scenarios of iron dose for hydrogen sulfide and struvite control

		Model		
Parameter	Data	without iron	with iron for H$_2$S control	with iron for struvite control
Iron dose, g FeCl$_3$/kg VSS		0	25	100
Influent total nitrogen, kg/d	306	308		
Influent total phosphorus, kg/d	90	89		
Influent total sulfur, kg/d	23	27		
Influent Ca, kg/d	127	128		
Influent Mg, kg/d	17	17		
Influent total solids	4.0%	4.0%		
Influent VSS, % of total solids	84%	85%		
Digester total solids	1.70%	2.10%	2.15%	2.24%
Digester VSS, % of total solids	67%	66%	65%	62%
Digester pH	7.15	7.16	7.07	6.80
Alkalinity as CaCO$_3$, mg/L	4,100	3,544	2,939	1,762
Gas production, Nm^3/h	176	172	174	178
Digester gas H$_2$S, ppm	**2,125**	**1,926**	**141**	**4**
Digester ammonia, mg N/L	1,169	979	978	1,018
Digester phosphate, mg P/L	156	166	113	2
Struvite, mg TSS/L	**?**	**842**	**827**	**37**
Vivianite, mg TSS/L		0	578	2,488
Iron sulfide, mg TSS/L		0	146	163
Elemental sulfur, mg S/L		0	3	1
Amorphous calcium phosphate, mg TSS/L		300	186	2

VSS: volatile suspended solids; TSS: total suspended solids.
Bold text refers to parameters controlled with iron dosing.

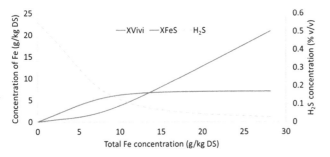

Figure 5 | Iron precipitates in function of iron dose and impact on H$_2$S concentration in the biogas.

as expected by the experts with a dose of 25 g $FeCl_3$/kg VSS, by precipitating sulfide into FeS. This process requires first the reduction of Fe^{3+} into Fe^{2+}, which is also correctly predicted. Furthermore, the 100 g $FeCl_3$/kg VSS dosage is shown to reduce drastically the struvite precipitation (from 842 mg TSS/L to 37 mg TSS/L), while increasing vivianite precipitation and decreasing phosphate concentration in the digester.

DISCUSSION

The application of the whole plant model behaviour for collection system, mainstream and sludge line enable evaluation of the new processes implemented. The modelling exercise on a sewer segment shows the potential of the model to simulate the behaviour of sulfur species and to evaluate the impact of dosing treatments in a segment of sewer system. It involves biological reduction of sulfate when no dosing is applied. The simulation of nitrate dosage shows the behaviour of chemical and biological sulfide oxidation processes and the kinetically limited oxidation of elemental sulfur. However, those simulations doe not take the biofilm in the sewer into account and thus neglect substantially the biological reactions that take place in the collection system. An adequate process unit model considering the biofilm is required as further development.

The application of the model on a typical mainstream configuration shows the impact of the sulfide oxidation on OUR and emphasizes the need to take into account sulfur reactions to appropriately model the oxygen demand depending on the influent sulfur concentrations and the length and characteristics of the sewer network. Furthermore, in the case of carbon source limitation, the sulfur-reducing organisms will be directly competing with the phosphorus accumulating organisms, reducing the biological phosphorus removal, which is not illustrated with this example.

The interconnections of phosphorus, iron and sulfur cycles are better demonstrated with the digester modelling. The role of sulfate-reducing bacteria in the production of hydrogen sulfide is shown by the model, which manages to predict the biogas quality accurately. The effect of iron dose on the biogas quality and precipitates is appropriately described, which gives confidence in the physical–chemical model. The precipitation processes are all implemented with the Koutsoukos *et al.* (1980) kinetic rate expression type, meaning that the precipitation rate depends on the distance to the thermodynamical equilibrium of the precipitate and on a kinetic rate parameter. The competition between different precipitates for the same ions will thus depend on the thermodynamic constants and on the kinetic rate parameters that must be calibrated. Both case studies show the ability of the model to predict the sequence of precipitates: in the presence of sulfur, iron dosage will first lead to iron sulfide precipitation, then to vivianite precipitation. The vivianite precipitation will thus compete with other phosphate precipitates, and especially with struvite as illustrated in Table 1.

CONCLUSIONS

Sulfur chemistry and biology occurring in activated sludge treatment, anaerobic digestion and sewers were introduced: (i) reduced and oxidized S species within the weak acid-base chemistry framework; (ii) biological oxidation/reduction, including the S_0 step; (iii) chemical redox reactions (sulfur oxidizes/reduces abiotically as well); (iv) H_2S gas transfer; (v) Fe^{2+} and Fe^{3+} redox transformations to be able to predict S bound with Fe in the digester; and (vi) extended Fe^{2+} and Fe^{3+} chemistry (hydroxides, HFO and vivianite formation).

The model can be used to develop strategies to cope with hydrogen sulfide production and optimize iron and nitrate addition in sewers, predict vivianite build-up in digesters, calculate oxygen demand of hydrogen sulfide in anaerobic wastewaters or return liquids, and correct pH predictions as hydrogen sulfide is a weak acid and sulfate is a strong acid.

SUPPLEMENTARY MATERIAL

The full Gujer matrix and parameters of the model can be downloaded at the following link: http://www.dynamita.com/public/models/Sumo2S.xlsm.

REFERENCES

Barrera, E. L., Spanjers, H., Solon, K., Amerlinck, Y., Nopens, I. & Dewulf, J. 2015 Modeling the anaerobic digestion of cane-molasses vinasse: extension of the anaerobic digestion model no. 1 (ADM1) with sulfate reduction for a very high strength and sulfate rich wastewater. *Water Research* **71**, 42–54.

Batstone, D. J. 2006 Mathematical modelling of anaerobic reactors treating domestic wastewater: rational criteria for model use.

Reviews in Environmental Science and Bio/Technology **5** (1), 57–71.

Batstone, D. J., Puyol, D., Flores-Alsina, X. & Rodríguez, J. 2015 Mathematical modelling of anaerobic digestion processes: applications and future needs. *Rev. Environ. Sci. Biotechnol.* **14**, 595–613.

Buisman, C., Uspeert, P., Janssen, A. & Lettinga, G. 1990 Kinetics of chemical and biological sulphide oxidation in aqueous solutions. *Water Research* **24** (5), 667–671.

Buisman, C., Ijspeert, P., Hof, A., Janssen, A., Tenhagen, R. & Lettinga, G. 1991 Kinetic parameters of a mixed culture oxidizing sulfide and sulfur with oxygen. *Biotechnology and Bioengineering* **38** (8), 813–820.

Cheng, X., Chen, B., Cui, Y., Sun, D. & Wang, X. 2015 Iron(III) reduction-induced phosphate precipitation during anaerobic digestion of waste activated sludge. *Separation and Purification Technology* **143**, 6–11.

Chou, H.-H., Huang, J.-S., Chen, W.-G. & Ohara, R. 2008 Competitive reaction kinetics of sulfate-reducing bacteria and methanogenic bacteria in anaerobic filters. *Bioresource Technology* **99** (17), 8061–8067.

Díaz, I. & Fdz-Polanco, M. 2012 Robustness of the microaerobic removal of hydrogen sulfide from biogas. *Water Science and Technology* **65**, 1368–1374.

Dynamita 2016 *Sumo User Manual*.

Dynamita 2018 http://www.dynamita.com/wp-content/uploads/Sumo2S.xlsm.

Fedorovich, V., Lens, P. & Kalyuzhnyi, S. 2003 Extension of anaerobic digestion model no. 1 with processes of sulfate reduction. *Applied Biochemistry and Biotechnology* **109** (1–3), 33–46.

Firer, D., Friedler, E. & Lahav, O. 2008 Control of sulfide in sewer systems by dosage of iron salts: comparison between theoretical and experimental results, and practical implications. *Science of the Total Environment* **392** (1), 145–156.

Flores-Alsina, X., Solon, K., Kazadi Mbamba, C., Tait, S., Gernaey, K. V., Jeppsson, U. & Batstone, D. J. 2016 Modelling phosphorus (P), sulfur (S) and iron (Fe) interactions for dynamic simulations of anaerobic digestion processes. *Water Research* **95**, 370–382. https://doi.org/10.1016/j.watres.2016.03.012.

Ge, H., Zhang, L., Batstone, D. J., Keller, J. & Yuan, Z. 2013 Impact of iron salt dosage to sewers on downstream anaerobic sludge digesters: sulfide control and methane production. *Journal of Environmental Engineering* **139** (4), 594–601.

Grau, P., Copp, J., Vanrolleghem, P. A., Takács, I. & Ayesa, E. 2009 A comparative analysis of different approaches for integrated WWTP modelling. *Water Science and Technology* **59**, 141–147. https://doi.org/10.2166/wst.2009.589.

Gutierrez, O., Park, D., Sharma, K. R. & Yuan, Z. 2010 Iron salts dosage for sulfide control in sewers induces chemical phosphorus removal during wastewater treatment. *Water Research* **44** (11), 3467–3475.

Hao, T., Xiang, P., Mackey, H. R., Chi, K., Lu, H., Chui, H., van Loosdrecht, M. C. M. & Chen, G.-H. 2014 A review of biological sulfate conversions in wastewater treatment. *Water Research* **65**, 1–21.

Hauduc, H., Takács, I., Smith, S., Szabo, A., Murthy, S., Daigger, G. T. & Spérandio, M. 2015 A dynamic physicochemical model for chemical phosphorus removal. *Water Research* **73**, 157–170.

Hvitved-Jacobsen, T., Vollertsen, J. & Nielsen, A. H. 2013 *Sewer Processes: Microbial and Chemical Process Engineering of Sewer Networks*, 2nd edn. CRC Press, Boca Raton, FL, USA.

Jenicek, P., Keclik, F., Maca, J. & Bindzar, J. 2008 Use of microaerobic conditions for the improvement of anaerobic digestion of solid wastes. *Water Science and Technology* **58**, 1491–1496.

Jiang, G., Sharma, K. R., Guisasola, A., Keller, J. & Yuan, Z. 2009 Sulfur transformation in rising main sewers receiving nitrate dosage. *Water Research* **43** (17), 4430–4440.

Kalyuzhnyi, S. V. & Fedorovich, V. V. 1998 Mathematical modelling of competition between sulphate reduction and methanogenesis in anaerobic reactors. *Bioresource Technology* **65** (3), 227–242.

Klok, J. B. M., de Graaff, M., van den Bosch, P. L. F., Boelee, N. C., Keesman, K. J. & Janssen, A. J. H. 2013 A physiologically based kinetic model for bacterial sulfide oxidation. *Water Research* **47** (2), 483–492.

Knobel, A. N. & Lewis, A. E. 2002 A mathematical model of a high sulphate wastewater anaerobic treatment system. *Water Research* **36** (1), 257–265.

Koutsoukos, P., Amjad, Z., Tomson, M. B. & Nancollas, G. H. 1980 Crystallization of calcium phosphates. A constant composition study. *Journal of the American Chemical Society* **102** (5), 1553–1557.

Liu, Y., Zhang, Y. & Ni, B.-J. 2015 Evaluating enhanced sulfate reduction and optimized volatile fatty acids (VFA) composition in anaerobic reactor by Fe (III) addition. *Environmental Science & Technology* **49** (4), 2123–2131.

Liu, Y., Tugtas, A. E., Sharma, K. R., Ni, B.-J. & Yuan, Z. 2016 Sulfide and methane production in sewer sediments: field survey and model evaluation. *Water Research* **89**, 142–150.

Lovley, D. & Phillips, E. 1988 Novel mode of microbial energy-metabolism – organic-carbon oxidation coupled to dissimilatory reduction of iron or manganese. *Applied and Environmental Microbiology* **54** (6), 1472–1480.

Luther, G. W., Findlay, A. J., MacDonald, D. J., Owings, S. M., Hanson, T. E., Beinart, R. A. & Girguis, P. R. 2011 Thermodynamics and kinetics of sulfide oxidation by oxygen: a look at inorganically controlled reactions and biologically mediated processes in the environment. *Frontiers in Microbiology* **2**, 62.

Mannucci, A., Munz, G., Mori, G. & Lubello, C. 2012 Biomass accumulation modelling in a highly loaded biotrickling filter for hydrogen sulphide removal. *Chemosphere* **88**, 712–717.

Nielsen, A. H., Vollertsen, J. & Hvitved-Jacobsen, T. 2003 Determination of kinetics and stoichiometry of chemical sulfide oxidation in wastewater of sewer networks. *Environmental Science & Technology* **37** (17), 3853–3858.

Nielsen, A. H., Lens, P., Vollertsen, J. & Hvitved-Jacobsen, T. 2005 Sulfide–iron interactions in domestic wastewater from a gravity sewer. *Water Research* **39** (12), 2747–2755.

Parker, C. D. 1945 The corrosion of concrete. I. The isolation of a species of bacterium associated with corrosion of concrete exposed to atmospheres containing hydrogen sulphide. *Experimental Biology and Medicine Journal* **23**, 81–90.

Poinapen, J. & Ekama, G. A. 2010 Biological sulphate reduction with primary sewage sludge in an upflow anaerobic sludge bed reactor – part 5: steady-state model. *Water SA* **36** (3), 193–202.

Roussel, J. & Carliell-Marquet, C. 2016 Significance of vivianite precipitation on the mobility of iron in anaerobically digested sludge. *Frontiers in Environmental Science* **4** (60), 1–12.

Solon, K., Flores-Alsina, X., Kazadi Mbamba, C., Ikumi, D., Volcke, E. I. P., Vaneeckhaute, C., Ekama, G., Vanrolleghem, P. A., Batstone, D. J., Gernaey, K. V. & Jeppsson, U. 2017 Plant-wide modelling of phosphorus transformations in wastewater treatment systems: impacts of control and operational strategies. *Water Research* **113**, 97–110. https://doi.org/10.1016/j.watres.2017.02.007.

Tichy, R., Janssen, A., Grotenhuis, J. T. C., Van Abswoude, R. & Lettinga, G. 1998 Oxidation of biologically-produced sulphur in a continuous mixed-suspension reactor. *Water Research* **32** (3), 701–710.

Utgikar, V. P., Harmon, S. M., Chaudhary, N., Tabak, H. H., Govind, R. & Haines, J. R. 2002 Inhibition of sulfate-reducing bacteria by metal sulfide formation in bioremediation of acid mine drainage. *Environ. Toxicol.* **17**, 40–48.

Varga, E., Hauduc, H., Barnard, J., Dunlap, P., Jimenez, J., Menniti, A., Schauer, P., Lopez-Vazquez, C. M., Gu, A. Z., Sperandio, M. & Takács, I. 2018 Recent advances in bio-P modelling – a new approach verified by full-scale observations. In: *6th IWA/WEF Water Resource Recovery Modelling Seminar*, 10–14 March 2018, *Lac Beauport, QC, Canada*.

WERF 2007 Minimization of odors and corrosion in collection systems phase1 – 04-CTS-1. WERF, Alexandria, VA, USA.

First received 29 June 2018; accepted in revised form 9 November 2018. Available online 21 November 2018

Presentation and evaluation of the zero-dimensional biofilm model 0DBFM

Mario Plattes

Mario Plattes
Luxembourg Institute of Science and Technology
 (LIST),
Environmental Research and Innovation (ERIN),
Site de Belvaux, 41 rue Brill, L-4422 Belvaux,
Luxembourg
E-mail: *mario.plattes@list.lu*

ABSTRACT

A zero-dimensional biofilm model, i.e. 0DBFM, has been developed for dynamic simulation of moving bed bioreactors (MBBRs). This mini-review aims at presenting and evaluating 0DBFM. 0DBFM is presented in Petersen matrix format and is based on the activated sludge model ASM1, which is an explicit and quite complex model (eight processes, 13 state variables, and 19 parameters) that has found wide application in engineering practice. 0DBFM is thus based on existing knowledge in biological wastewater treatment. The ASM1 approach has been confirmed by respirometry since the resulting respirograms were successfully simulated with ASM1. 0DBFM distinguishes between attached and suspended biomass and incorporates attachment of suspended matter from the bulk liquid onto the biofilm and detachment of biofilm into the bulk liquid. Still, 0DBFM respects the golden rule of modelling, which says that 'models should be as simple as possible and as complex as needed' and resists Occam's razor. The practicability of 0DBFM has been shown on a pilot-scale plant since nine days of wastewater treatment were successfully simulated and effluent quality was dynamically predicted. Finally, 0DBFM can be inspiring and the applicability of 0DBFM to other biofilm systems can be tested.

Key words | biofilm, modelling, Occam's razor, simulation, wastewater, zero-dimensional biofilm
model

INTRODUCTION

The water resource recovery facility (WRRF) of Hesperange in Luxemburg was modernised and upgraded with moving bed bioreactor (MBBR) technology. In this context, a scientific support with emphasis on process modelling and simulation was carried out.

Note that a gap between biofilm research and engineering practice has increased over the past decades in the biofilm modelling community. The reasons for this have been discussed in the literature (Noguera *et al.* 1999; Morgenroth *et al.* 2000a). Considerable research has therefore been dedicated to the topic and an excellent overview of available biofilm models can be found in the *IWA Scientific and Technical Report No. 18* (Eberl *et al.* 2006). However, there is still a need to develop biofilm models that can be used in engineering practice. In particular, there is a need for biofilm models that can predict WRRF effluent quality in response to influent variations. One-dimensional (1D) biofilm models have been mostly used for this purpose since there was a trend from more complex two- and three-

dimensional (2D and 3D) models towards 1D models for application in engineering practice (Morgenroth *et al.* 2000b). Although 1D models have been proposed since the 1970s (Wuertz & Falkentoft 2003), calibration protocols for 1D models are still under development (see for example Barry *et al.* 2012 and Rittmann *et al.* 2018) and still need to be established in the engineering community. This illustrates how difficult the application of even relatively simple 1D biofilm models is in engineering practice. After the author of this contribution had a negative experience with a 1D biofilm model, the objective of the research presented here became to develop a biofilm model for engineering practice.

It was observed that the biofilm, which was attached to the carriers in a pilot-scale plant, had a complex three-dimensional structure with cell clusters, pores and channels, not in line with the simple schematic one-dimensional representation of a biofilm but more of an activated sludge matrix. The idea came to model the MBBR with an activated sludge model, i.e. ASM1 (Henze *et al.* 2000). This approach

doi: 10.2166/wst.2018.450

resulted in 0DBFM, a zero-dimensional (0D) biofilm model that can dynamically predict effluent quality in response to influent variations and that is presented and evaluated here. A brief account of the methods that were employed is given in order to facilitate understanding of what has been done.

METHODS

The methods regarding wastewater analysis, operation of the pilot-scale plant, and respirometry are described in detail elsewhere (Plattes *et al.* 2006, 2007). A brief account of what has been done is given here. The pilot-scale MBBR was composed of three subsequent compartments. The first and second compartment (2.8 m³ each) were filled with Kaldnes (K1) carrier elements with a filling degree of 50% and 65% respectively. The third compartment was used as a settling tank. During the first measurement campaign (5 d) the first and the second compartment were both aerated, i.e. nitrification mode. The inflow was 21.4 m³/d of raw municipal wastewater. During the second measurement campaign (4 d) the first compartment was anoxic, the second compartment was aerated, and nitrates were recycled from the settling tank to the first compartment, i.e. denitrification mode. The inflow was 17.05 m³/d and the nitrate recycle was 22.25 m³/d (recycle ratio 1.3). The MBBR had attained stable operation before the measurement campaigns took place. During the first measurement campaign (nitrification mode) 6 h composite samples of influent and effluent were analysed for standard wastewater parameters. During the second measurement campaign (denitrification mode) 4 h composite samples were analysed for the same parameters and respirometric experiments were carried out according to the static gas static liquid principle (Spanjers *et al.* 1998). The respirograms were simulated with the WRRF simulator GPS-X (Hydromantis Inc.) using ASM1 (Henze *et al.* 2000). The resulting kinetic parameters were transferred to the proposed zero-dimensional biofilm model, that was also implemented in GPS-X. The amount of biofilm (biofilm solids) and the amount of suspended solids in the reactor were also measured.

PRESENTATION OF 0DBFM

The mass balances of 0DBFM are analogous to the mass balances in activated sludge models, with the only difference being that biofilm remains in the reactor: the rate of change of dissolved state variables (S_i) with respect to time (t) equals the hydraulic flow (Q) divided by the reactor volume (V) multiplied by the difference between the concentration in the influent ($S_{i,in}$) and the concentration in the effluent ($S_{i,out}$) plus the rate in the zero-dimensional biofilm model (r_{0DBFM}) (Equation (1)). The rate of change of suspended particulates (X_i^S) (Equation (2)) is analogous to Equation (1). The rate of change of attached (biofilm) particulates (X_i^B) is simply given by the rate of change of 0DBFM (r_{0DBFM}), because biofilm does not enter or leave the reactor with the hydraulic flow (Equation (3)).

$$\frac{dS_i}{dt} = \frac{Q}{V}(S_{i,in} - S_{i,out}) + r_{0DBFM} \tag{1}$$

$$\frac{dX_i^S}{dt} = \frac{Q}{V}(X_{i,in}^S - X_{i,out}^S) + r_{0DBFM} \tag{2}$$

$$\frac{dX_i^B}{dt} = r_{0DBFM} \tag{3}$$

0DBFM has been presented in Petersen matrix format and contains 29 processes, 21 state variables and 28 parameters, and is presented here as published in the literature (Plattes *et al.* 2008). The biochemical conversions in 0DBFM are as given in ASM1. They apply to attached and suspended biomass and are a function of the concentration of substrates in the bulk liquid. Diffusional mass transfer limitations manifest by adapted half-saturation indices in the Monod terms of 0DBFM.

Further, 0DBFM has attachment of suspended solids from the bulk liquid onto the biofilm and detachment of biofilm into the bulk liquid. Attachment rate (r_a) and detachment rate (r_d) are formulated according to a detachment rate expression proposed in the literature (Trulear & Characklis 1982). It is assumed that the attachment rate is proportional to the square of the amount of suspended solids (SS) and that the detachment rate is proportional to the square of the amount of biofilm solids (BS), the proportionality factor being the attachment rate constant (k_a) and the detachment rate constant (k_d) respectively (Equations (4) and (5)). Attachment and detachment rates are formulated individually for each particulate state variable in 0DBFM, whilst the values of k_a and k_d are maintained as constant (see Plattes *et al.* 2008 for further explanation).

$$r_a = k_a \cdot (SS)^2 \tag{4}$$

$$r_d = k_d \cdot (BS)^2 \tag{5}$$

The model therefore distinguishes between hydraulic residence time (HRT) and residence time of biofilm on the substratum, which is called the biofilm age (BA). Biofilm age can be estimated using simulation results and Equation (6). Biofilm age can be estimated only, because the detachment rate is not measurable.

$$BA = \frac{\text{Amount of biofilm}}{\text{Detachment rate}} \qquad (6)$$

0DBFM does not incorporate biofilm structure in any form and therefore does not contain any parameter related to biofilm structure (like biofilm thickness).

Kinetic parameters are obtained from respirometry, if available, for model calibration. Otherwise, calibration can be started with default values of ASM1. Further model calibration is achieved using two iterative steps:

Step 1: The attachment and detachment rate constants are adjusted in order to match the biofilm solids in the compartment.

Step 2: The detachment rate constant is further adjusted in order to produce the required biofilm age and nitrification rate using the effluent ammonium and/or nitrate concentration.

Step 2 changes the amount of biofilm (biofilm solids) and the attachment rate needs to be readjusted. Steps 1 and 2 become iterative in the procedure.

Kinetic parameters can be further adjusted using the effluent data. Note that half-saturation indices can be adjusted individually for attached and suspended biomass.

DISCUSSION AND EVALUATION OF 0DBFM

0DBFM is based on ASM1 (Henze *et al.* 2000) and hence the Monod model (Monod 1949). The Monod model gives a functional relation between specific growth rate and substrate concentration in the *bulk*. ASM1 and the Monod model have found wide application in engineering practice so far. 0DBFM is thus based on existing and well-established knowledge in biological wastewater treatment modelling and biochemical engineering.

In activated sludge models (ASMs) the Monod model is implemented to describe the macrokinetic behaviour of activated sludge reactors, not explicitly describing concentration gradients in the activated sludge floc (Plattes 2009). Diffusional mass transport limitations are taken into account implicitly by adapted values of the half-saturation indices in ASM1, which are a function of various processes in activated sludge systems (Arnaldos *et al.* 2015). So far, in biofilm modelling the Monod equation has been used to model intrinsic kinetics of the biofilm, since diffusion is usually described explicitly using Fick's laws of diffusion. In 0DBFM, the Monod model is used to describe the macrokinetic behaviour of biofilm reactors, analogous to the use of the Monod model in state of the art activated sludge models (ASMs). The role that biofilm and floc structure have played in this context has been thoroughly discussed in the literature (Plattes 2009). Essentially, it is stated that biofilm structure has been strongly emphasized in the biofilm modelling community (1D, 2D, 3D), whilst state of the art activated sludge models do not take structure into account, i.e. they are zero-dimensional (0D). The author believes that the application of Fick's laws of diffusion, which crucially link diffusional mass transfer to biofilm structure, has been a driving force for biofilm model development from 1D to 3D, in addition to the fact that biofilms have a complex three dimensional structure (see Wanner 1995 and references in there).

The zero-dimensional ASM1 approach of 0DBFM has been confirmed by respirometry in the laboratory (Plattes *et al.* 2007). The obtained respirograms featured the typical endogenous and exogenous respiration phases and the respirograms could be simulated with ASM1 (Figure 1). The resulting kinetic parameters were transferred to the biofilm model of the pilot-scale MBBR. The fact that the respirograms have the shape of typical respirograms obtained with activated sludge justifies the modelling approach taken by 0DBFM, i.e. modelling a biofilm system with an activated sludge model.

0DBFM has attachment and detachment of biomass, two processes that are in the eyes of the author not measurable, because they are counteractive and simultaneous. In order to avoid unnecessary complexity, attachment and detachment might therefore not be integrated in a biofilm model, although it is recognized that both processes take place in biofilm systems (Hermanowicz 2003), detachment being considered to be a process that is often overlooked (Morgenroth 2003). Indeed, a zero-dimensional biofilm model that ignores both attachment and detachment has been formulated and applied by other researchers (Volcke *et al.* 2008). However, by the author's own experience it was necessary to integrate attachment and detachment in 0DBFM, in order to distinguish between attached and suspended biomass, a prerequisite for the application of the proposed mass balances (Equations (2) and (3)) and the discrimination of hydraulic residence time and biofilm age

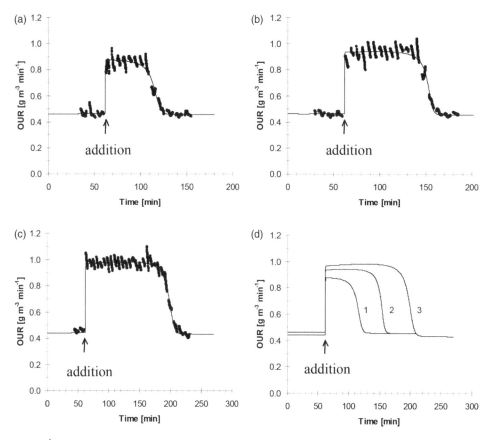

Figure 1 | Measured and simulated respirograms (adapted from Plattes *et al.* 2007) obtained with biofilm samples originating from the pilot-scale MBBR (addition of 5.25 (a), 10.49 (b), and 17.51 mg/l (c) ammonium-N; simulations compared in (d)).

(Equation (6)). A similar approach including attachment and detachment has been taken previously in a pure simulation study to model a continuous stirred tank biofilm reactor (Chen & Chai 2005). Further, the simulation results give at least a quantitative estimation of what the attachment and detachment rate possibly could be, which is interesting since real data regarding these rates are scarce or non-existent. Note that estimation of attachment and detachment rate is also possible with 1D biofilm models.

The proposed calibration procedure is relatively simple because no structure related parameters (like biofilm thickness) need to be calibrated.

The practicability of 0DBFM has been demonstrated on the pilot-scale plant: nine days of wastewater treatment were successfully simulated and the dynamic variations of ammonium and nitrate nitrogen were accurately predicted by 0DBFM (see Figure 2 for a five day simulation result).

Although 0DBFM is quite complex and explicit it respects the golden rule of modelling, which says that 'models should be as simple as possible and as complex as

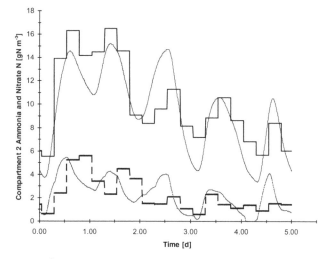

Figure 2 | Simulated (waved) and measured (stepped) ammonium (bottom) and nitrate (top) nitrogen concentration in the effluent of the pilot-scale MBBR operated in nitrification mode (adapted from Plattes *et al.* 2008).

needed'. Further 0DBFM conforms to the principle of parsimony and resists Occam's razor in contrast to 1D, 2D and 3D biofilm models.

Finally, the zero-dimensional biofilm modelling approach can be inspiring and the applicability to other wastewater treatment systems like fixed bed reactors and granular sludge reactors, for example, could be tested.

CONCLUSIONS

A fundamental problem has been overcome by 0DBFM: modelling biofilm structure. It is therefore believed that 0DBFM can possibly help in reducing the gap that has developed over the past decades between biofilm research and engineering practice in the biofilm modelling community. The predictive power of 0DBFM should be further tested on pilot- and full-scale plants and the simulation results should be compared to 1D models. Long term model performance should be evaluated.

ACKNOWLEDGEMENTS

The author acknowledges funding of the project FNR/01/03/06 by the 'Fond National de la recherche' (Luxembourg).

REFERENCES

Arnaldos, M., Amerlinck, Y., Rehman, U., Maere, T., Van Hoey, S., Naessens, W. & Nopens, I. 2015 From the affinity constant to the half-saturation index: understanding conventional modelling concepts in novel wastewater treatment processes. *Water Research* **70**, 458–470.

Barry, U., Choubert, J. M., Canler, J. P., Héduit, H., Robin, L. & Lessard, P. 2012 A calibration protocol of a one-dimensional moving bed bioreactor (MBBR) dynamic model for nitrogen removal. *Water Science and Technology* **65** (7), 1172–1178.

Chen, L.-M. & Chai, L.-H. 2005 Mathematical model and mechanisms for biofilm wastewater treatment systems. *World Journal of Microbiology & Biotechnology* **21**, 1455–1460.

Eberl, H., Morgenroth, E., Noguera, D. R., Picioreanu, C., Rittmann, B., Van Loosdrecht, M. & Wanner, O. 2006 *Mathematical Modelling of Biofilms* (IWA Task Group on Biofilm Modelling, ed.). Scientific and Technical Report No. 18, IWA Publishing, London, UK.

Henze, M., Gujer, W., Mino, T., Van Loosdrecht, M. C. M. 2000 *Activated Sludge Models ASM1, ASM2, ASM2d and ASM3* (IWA Task Group on Mathematical Modelling for Design and Operation of Biological Wastewater Treatment, ed.). Scientific and Technical Report No. 9, IWA Publishing, London, UK.

Hermanowicz, S. W. 2003 Biofilm architecture: interplay of models and experiments. In: *Biofilms in Wastewater Treatment – An Interdisciplinary Approach* (S. Wuertz, P. L. Bishop & P. A. Wilderer, eds). IWA Publishing, London, UK, pp. 32–48.

Monod, J. 1949 The growth of bacterial cultures. *Annual Review of Microbiology* **3**, 371–394.

Morgenroth, E. 2003 Detachment: an often-overlooked phenomenon in biofilm research and modeling. In: *Biofilms in Wastewater Treatment – An Interdisciplinary Approach* (S. Wuertz, P. L. Bishop & P. A. Wilderer, eds). IWA Publishing, London, UK, pp. 264–288.

Morgenroth, E., Van Loosdrecht, M. C. M. & Wanner, O. 2000a Biofilm models for the practitioner. *Water Science and Technology* **41** (4–5), 509–512.

Morgenroth, E., Eberl, H. & Van Loosdrecht, M. C. M. 2000b Evaluating 3-D and 1-D mathematical models for mass transport in heterogeneous biofilms. *Water Science and Technology* **41** (4–5), 347–356.

Noguera, D. R., Okabe, S. & Picioreanu, C. 1999 Biofilm modelling: present status and future directions. *Water Science and Technology* **39** (7), 273–278.

Plattes, M. 2009 The Role of Biofilm and Floc Structure in Biological Wastewater Treatment Modelling. In: *Biochemical Engineering* (F. E. Dumont & J. A. Sacco, eds). Nova Science Publishers, New York, USA, pp. 245–255.

Plattes, M., Henry, E., Schosseler, P. M. & Weidenhaupt, A. 2006 Modelling and dynamic simulation of a moving bed bioreactor for the treatment of municipal wastewater. *Biochemical Engineering Journal* **32**, 61–68.

Plattes, M., Fiorelli, D., Gillé, S., Girard, C., Henry, E., Minette, F., O'Nagy, O. & Schosseler, P. M. 2007 Modelling and dynamic simulation of a moving bed bioreactor using respirometry for the estimation of kinetic parameters. *Biochemical Engineering Journal* **33**, 253–259.

Plattes, M., Henry, E. & Schosseler, P. M. 2008 A zero-dimensional biofilm model for dynamic simulation of moving bed bioreactor systems: model concepts, Peterson matrix, and application to a pilot-scale plant. *Biochemical Engineering Journal* **40**, 392–398.

Rittmann, B. E., Boltz, J. P., Brockmann, D., Daigger, G. T., Morgenroth, E., Sørensen, K. H., Takács, I., van Loosdrecht, M. & Vanrolleghem, P. 2018 A framework for good biofilm reactor modeling practice (GBRMP). *Water Science and Technology* **77** (5), 1149–1164.

Spanjers, H., Vanrolleghem, P. A., Olsson, G. & Dold, P. 1998 *Respirometry in Control of the Activated Sludge Process: Principles*. IAWQ Task Group on Respirometry, Scientific and Technical Report No. 7, IAWQ Publishing, London, UK.

Trulear, M. G. & Characklis, W. G. 1982 Dynamics of biofilm processes. *Journal of the Water Pollution Control Federation* **54**, 1288–1301.

Volcke, E. I. P., Sanchez, O., Steyer, J.-P., Dabert, P. & Bernet, N. 2008 Microbial population dynamics in nitrifying reactors: experimental evidence explained by a simple model including interspecies competition. *Process Biochemistry* **43**, 1398–1406.

Wanner, O. 1995 New experimental findings and biofilm modelling concepts. *Water Science and Technology* **32** (8), 133–140.

Wuertz, S. & Falkentoft, C. M. 2003 Modelling and simulation: Introduction. In: *Biofilms in Wastewater Treatment – An Interdisciplinary Approach* (S. Wuertz, P. L. Bishop & P. A. Wilderer, eds). IWA Publishing, London, UK, pp. 3–7.

First received 24 April 2018; accepted in revised form 16 October 2018. Available online 24 October 2018

Modeling quaternary ammonium compound inhibition of biological nutrient removal activated sludge

Daniela Conidi, Mehran Andalib, Christopher Andres, Christopher Bye, Art Umble and Peter Dold

ABSTRACT

Quaternary ammonium compounds (QACs) are surface-active organic compounds common in industrial cleaner formulations widely used in various sanitation applications. While acting as effective pathogenic biocides, QACs lack selective toxicity and often have poor target specificity. As a result, adverse effects on biological processes and thus the performance of biological nutrient removal (BNR) systems may be encountered when QACs enter wastewater treatment plants (WWTPs). Because of these impacts, there is motivation to screen wastewater influents for QACs and for process engineers to consider the inhibition effects of QACs on process evaluation and design of BNR plants. This paper introduces a mathematical model to describe the fate of QACs in a WWTP via biodegradation and bio-adsorption, and the inhibitory effect of QACs on nitrifiers and ordinary heterotrophic organisms. The model was incorporated as an add-on model in BioWin 5.3 and simulations of experimental systems were used for comparison of model results to measured data reported in the literature. The model was found to accurately predict the bulk phase concentration of QAC and the inhibition of nitrification with QAC concentrations ≥ 2 mg/L. This work provides a preliminary framework for simulation of BNR plants receiving inhibitory substances in the influent.

Key words | BNR, inhibition, modeling, quaternary ammonium compounds

Daniela Conidi
Christopher Bye
Peter Dold
Envirosim Associates Ltd,
175 Longwood Rd S, Suite 114A, Hamilton, ON L8P
  0A1,
Canada

Mehran Andalib (corresponding author)
Christopher Andres
Art Umble
Stantec Inc.,
Edmonton, AB T5 K 2L6,
Canada
E-mail: *mehran.andalib@stantec.com*

INTRODUCTION

Quaternary ammonium compounds (QACs) are used extensively in domestic, agricultural, health care, and industrial applications as surfactants, emulsifiers, fabric softeners, disinfectants, pesticides, corrosion inhibitors, paint additives, cosmetics and personal care products (Yang *et al.* 2014). In 2004, global annual consumption of QACs was estimated at 500,000 tons and increasing (Chen *et al.* 2018). The widespread use of QACs means that they may be present in many wastewater treatment plant (WWTP) influents. It has been reported that roughly 75% of all QACs consumed end up in influent to WWTPs (Carbajo *et al.* 2015). QAC concentrations in the range of 25–300 mg/L, 0.3–3.6 mg/L and 22–103 mg/kg have been reported in the influents, effluents, and sludges of WWTPs, respectively (Ruan *et al.* 2014; Carbajo *et al.* 2015; Khan *et al.* 2015; Zhang *et al.* 2015). In addition, mean concentrations of 3,700 mg/kg have been reported in the sludge from five WWTPs in Switzerland (Zhang *et al.* 2015).

QACs are composed of at least one hydrophobic hydrocarbon chain linked to a positively charged nitrogen atom, and other alkyl groups which are mostly short-chain substituents such as methyl or benzyl groups. This structure gives them unique physical and chemical properties (Ren *et al.* 2011). While acting as effective biocides against a wide range of pathogenic microorganisms, QACs lack selective toxicity and often have poor target specificity. As a result, they negatively impact the physiological groups responsible for wastewater treatment and thus the performance of biological nutrient removal (BNR) systems. For example, QACs have been found to inhibit respiratory enzymes decreasing the rate of chemical oxygen demand (COD) substrate utilization (Zhang *et al.* 2010). QACs also have an adverse effect on nitrification. Total inhibition of nitrification was found at a QAC concentration of 2 mg/L due to enzymic toxicity (Sütterlin *et al.* 2008), while a QAC concentration greater than 50 mg/L has been reported to inhibit

doi: 10.2166/wst.2018.449

heterotrophic denitrification, with low temperatures exacerbating the inhibitory effect on nitrite reduction (Hajaya 2011; Yang *et al.* 2014). In addition, a QAC concentration of 50 mg/L was found to affect anaerobic degradation resulting in reduced methane production and volatile fatty acid accumulation (Tezel 2009).

It has been reported in the literature that the most effective method for removal of QACs in a treatment facility is through adsorption processes, including adsorption to activated sludge biomass (Ren *et al.* 2011). The tendency for QACs to adsorb onto solids and accumulate in the WWTP has been shown to increase with increased alkyl chain length. QAC sorption also is strongly related to temperature, with decreasing temperature resulting in an increased sorption rate onto activated sludge (Zhang *et al.* 2015). Equilibrium partitioning data between the solid and liquid phase have been described well by both Langmuir and Freundlich isotherm models (Ren *et al.* 2011). Adsorption kinetics have been best described by a pseudo-second-order model (Ren *et al.* 2011). Higher initial concentrations in the bulk liquid increase the adsorption capacity (Chen *et al.* 2018).

Another QAC removal mechanism is biodegradation. QACs are considered to be biodegradable upon complete depletion of available readily and slowly biodegradable COD (Zhang *et al.* 2010). Since QACs also inhibit respiration and hence COD utilization, respiratory inhibition is also responsible for the fate of QACs in activated sludge. In addition, microbial acclimation and enrichment has been shown to contribute to reduced inhibition and enhanced biodegradation of QACs in laboratory-scale BNR systems (Hajaya & Pavlostathis 2012).

As a result of the negative impacts of QACs on biological treatment systems, there is motivation to screen wastewater influents for QACs and for process engineers to consider the inhibition effects of QACs on process evaluation and design of BNR plants. Designs should provide adequate mixed liquor levels for proper adsorption of QACs and biodegradation to below the nitrifier inhibitory threshold. The objectives of the present study are to:

(1) introduce a mathematical model to simulate the different degrees of QAC inhibition of nitrifiers (ammonia- and nitrite-oxidizers, AOB, NOBs) and ordinary heterotrophic microorganisms (OHOs) at different operational conditions;
(2) enhance the understanding of the fate and effect of QACs in engineered treatment systems, in turn contributing to the effective design and management of QAC-containing wastewaters; and

(3) provide an effective framework for proper simulation and design of a WWTP with inhibitory substances.

MATERIALS AND METHODOLOGY

Literature data

The present study leveraged data and observations from two key sources:

(1) Hajaya, M. 2011 *Fate and Effect of Quaternary Ammonium Antimicrobial Compounds on Biological Nitrogen Removal within High-Strength Wastewater Treatment Systems*. PhD Thesis, Georgia Institute of Technology, Atlanta, GA, USA.
(2) Yang, J., Tezel, U., Li, K. & Pavlostathis, S. G. 2014 Prolonged exposure of mixed aerobic cultures to low temperature and benzalkonium chloride affect the rate and extent of nitrification. *Bioresource Technology*, **179**, 193–201.

Hajaya (2011)

Hajaya assessed and quantified the inhibitory effect of QACs and evaluated the role of adsorption, inhibition, and biodegradation on the fate and effect of QACs in a BNR system treating poultry-processing wastewater. Figure 1 illustrates the continuous flow BNR system, which included pre-anoxic, anoxic, and aerobic reactors with volumes of 4, 4, and 5 L, respectively. The settler volume was 1.5 L. Reactors 1 and 2 were mechanically mixed while mixing in reactor 3 was accomplished via diffused aeration. Raw wastewater was fed at a rate of 2.0 L/day at 4 °C. The temperature in the process was maintained at 22 °C. Total design, actual anoxic, and actual aerobic solids retention times (SRTs) were 25, 10.4, and 12.6 days, respectively. The WAS and nitrate return flow rates were 0.35 and 4.0 L/day, respectively. Target MLSS was 1,200 mg/L and the mixed-liquor target pH was 7.0. The target dissolved oxygen (DO) in

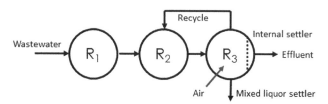

Figure 1 | Process flow diagram of simulated continuous-flow multi-stage BNR system.

the aerobic zone (reactor 3) was 3–5 mg/L, resulting in an airflow of 3–6 m^3/h.

The influent poultry processing wastewater TSS, VSS, TCOD, sCOD, NH$_3$-N, and TN concentrations were 125 mg/L, 120 mg/L, 1,275 mg/L, 919 mg/L, 46 mg N/L and 103 mg N/L, respectively. A blend of three benzalkonium chloride (BAC) (the most common QAC found in wastewater) homologs were used in the Hajaya study as follows: 32% C12BAC/40% C14BAC/8% C16BAC/10% ethanol/10% water. The BNR system was operated without any QAC addition for a period of 30 days, with full nitrification and significant denitrification. From days 30 to 80, a concentration of 5 mg/L QAC was injected continuously in the process.

It is noteworthy to mention that during QAC addition, the concentration of QAC in the pre-anoxic zone and aerobic zone did not exceed 0.3 mg/L, which is well below the established inhibitory threshold for nitrifiers. From day 30 to day 381 the concentration of QAC in the influent stream was increased to 60 mg/L in five incremental steps. During this stage, the QAC concentration in the bulk liquid of the aerobic zone increased from 0.3 mg/L to 1.8 mg/L, still below the nitrifier inhibitory threshold concentration in the bulk liquid. As expected, no inhibition of nitrification or denitrification rates was observed. Because QAC concentrations in the bulk liquid were less than 2 mg/L, Hajaya's study has limited value in terms of validating the proposed model's predicted inhibitory effects of QACs on nitrification. However, it does provide a meaningful dataset that can be used for comparison of predicted QAC adsorption and biodegradation.

Yang *et al.* (2014)

Yang *et al.* assessed (1) the effect of QAC concentration on nitrification at room temperature and (2) investigated the combined effects of QAC and prolonged exposure to low temperature on nitrification. This was accomplished by a series of nitrification assays. For the first objective, a series of short-term batch assays was conducted at room temperature with mixed nitrifying culture developed with mixed-liquor from the RM Clayton WWTP in Atlanta, GA, USA. The steady-state MLVSS was 290 mg VSS/L. Five 200 mL samples of the culture were collected, amended with 100 mg N/L NH$_4$Cl, and aliquots of QAC were added to each to generate initial QAC concentrations of 5, 10, 15, and 20 mg/L. Each of the five samples were continuously aerated and mixed. The concentration of nitrogen species and QACs were measured. Ammonia oxidation at room

temperature (22–24 °C) by a nitrifying culture was inhibited at increasing QAC concentrations. Significant nitrification inhibition was observed as low as 5 mg QAC/L and nitrification essentially ceased at 15 mg QAC/L. No QAC degradation was observed in the short-term assay (96 hours). Data from this part of the Yang *et al.* study were used to validate predicted QAC inhibition of nitrification.

Model development and assumptions

In this work, a mathematical model was developed to incorporate the following processes:

(1) adsorption of QAC onto biomass via bio-adsorption kinetics;
(2) inhibition of AOBs, NOBs, OHOs under aerobic and anoxic conditions at different liquid phase QACs thresholds; and
(3) biodegradation of liquid phase QACs by OHOs under aerobic and anoxic conditions.

The model assumes that microorganisms are already acclimated to QACs; no population shifts or changes in metabolic abilities were modelled with time. It is also assumed that only the liquid phase concentration of QAC was inhibitory (i.e. adsorbed QAC did not have any inhibition impacts).

To develop the biokinetic expressions describing the impact of QAC inhibition on organism (AOB, NOB, OHO) growth rate, various inhibition expressions from the literature were evaluated. The formulation of the inhibition expressions evaluated included an exponential inhibition term proposed in Aiba *et al.* (1968) (Equation (1)), a Monod-type inhibition term (Equation (2)), and a Haldane inhibition term (Equation (3)), where C_i is the concentration of the inhibitory compound (i.e. QAC) and K_i is the inhibition coefficient:

$$exp^{-\frac{Ci}{Ki}} \tag{1}$$

$$\frac{Ki}{Ci + Ki} \tag{2}$$

$$\frac{Ci}{K_s + Ci + \frac{Ci^2}{Ki}} \tag{3}$$

Based on comparisons to the experimental literature data reported by Hajaya (2011) and Yang *et al.* (2014), the most accurate inhibition model for AOBs and NOBs was found with a Monod-type inhibition expression. Aerobic and anoxic growth of OHOs on QAC (Ci) was found to be best described using Haldane substrate utilization kinetics. With Haldane kinetics, as the value of the inhibition parameter (Ki) decreases, the degree of inhibition on growth increases. The impact of temperature, limiting nutrient concentrations (e.g. ammonia, phosphate, CO_2, cations, anions) and pH inhibition were also included in the biokinetic expressions for organism growth.

Adsorption data from the literature were evaluated to determine the appropriate formulation of QAC adsorption onto biomass (i.e. bio-adsorption kinetics). Ren *et al.* (2011) found that equilibrium adsorption of QACs onto activated sludge at 25 °C was best described using a Langmuir isotherm. Adsorption experiments were carried out by Ren *et al.* (2011) with initial QAC concentrations in the range of 10–140 mg/L at 25 °C. The sludge concentration was 250 mg/L and the contact time was 4 h. Adsorption isotherms were also reported at different temperatures; adsorption of QAC onto activated sludge was found to be inversely proportional to temperature. The maximum sorbed QAC concentration (qmax, g/g) was extrapolated from the temperature dependent isotherms reported by Ren *et al.* (2011) and a relationship was developed to correct the maximum amount of QAC that can be sorbed for temperature. The following isotherm (Equation (4)) was derived to describe adsorption of QAC onto biomass:

$$q_e = qmax \times 0.988^{(T-15)} \cdot \frac{Kl \cdot C_e}{1 + Kl \cdot C_e} \qquad (4)$$

where q_e is the sorbed phase QAC concentration at equilibrium, qmax is the Langmuir isotherm constant (reported as 0.3683 g/g by Ren *et al.* (2011)), Kl is the Langmuir isotherm constant (reported as 0.047 L/mg by Ren *et al.* 2011), and C_e is the equilibrium concentration of QAC after 4 h. Ren *et al.* (2011) also carried out kinetic studies to characterize QAC adsorption over time. The kinetics of QAC adsorption onto activated sludge were found to be best described by Ren *et al.* (2011) with a pseudo-second-order kinetic expression (Equation (5)):

$$\frac{dq_t}{dt} = k(q_e - q_t)^2 \qquad (5)$$

where q_t is the sorbed phase QAC concentration at time t. The final adsorption rate equation used for this study (Equation (6)) combined the isotherm and second-order kinetic expressions described above accounting for the

appropriate unit conversions:

$$Rate_{ads} = \frac{1000}{TSS} \cdot k \cdot \left(qmax \times 0.988^{(T-15)} \cdot \frac{Kl \cdot C_e}{1 + Kl \cdot C_e} - \frac{C_{ads}}{TSS} \right)^2 \qquad (6)$$

where 1,000 is a units conversion constant (mg/g), C_{ads} is the concentration of adsorbed QAC, and k is the adsorption rate constant.

The overall mathematical model was incorporated as an add-on model in the BioWin 5.3 wastewater treatment process simulator using BioWin's Model Builder functionality. Table 1 outlines the processes and associated kinetic rate equations developed for this model. The concentration of liquid phase QAC is denoted as Ci. Table 2 outlines the associated stoichiometry matrix for each of the processes described in Table 1. The rate and stoichiometric constants are summarized in Table 3. The rate constants were based on data provided in the literature or derived via comparison of the model predictions to measured data in the literature.

RESULTS AND DISCUSSION

Both the acclimated bench-scale BNR system utilized by Hajaya (2011) and the non-acclimated batch system utilized by Yang *et al.* (2014) were configured and simulated in BioWin 5.3 using the reported operational parameters. The observations noted from experimentation were used for calibration of the model developed in this study, not for comparison between the literature studies.

Hajaya (2011) reported performance results from the BNR system during continuous operation with poultry processing wastewater containing QACs. Table 4 summarizes the measured *versus* predicted performance results from the last reactor (R3) and the effluent of the BNR system (see Figure 1). The predicted results were obtained via steady state simulations. The modeled parameters are within the same range as the measured values. There are some minor deviations; likely these are due to some uncertainly around the detailed wastewater characteristics. In addition, the parameters reported by Hajaya were averaged during continuous operation with slight variations at different QAC target concentrations in the influent while the simulated results were obtained with constant QAC target concentrations of 5, 10, 15, 45 and 60 mg QAC/L.

Hajaya (2011) reported the steady state QAC concentrations throughout the BNR system (Figure 1) during operation with step-increased influent QAC concentrations

Table 1 | Summary of process rate equations for all processes

No.	Biological process	Reaction rates
1	AOB growth with Ci inhibition	$MuMax_{AOB} \cdot ThetaMuaob^{T-20} \cdot \dfrac{DO}{K_{oaob}+DO} \cdot \dfrac{NH_3N}{K_{aob}+NH_3N} \cdot \dfrac{CO_2}{K_{sCO2}+CO_2} \cdot \dfrac{PO_4P}{P04P_{limit}+PO_4P} \cdot \dfrac{Kiaob}{Ci+Kiaob} \cdot Z_{AOB} \cdot pHinhibition$
2	NOB growth with Ci inhibition	$MuMax_{NOB} \cdot ThetaMunob^{T-20} \cdot \dfrac{DO}{K_{onob}+DO} \cdot \dfrac{NO_2\text{-}N}{K_{nO2}+NO_2\text{-}N} \cdot \dfrac{CO_2}{K_{CO2}+CO_2} \cdot \dfrac{PO_4P}{P04P_{limit}+PO_4P} \cdot \dfrac{Kinob}{Ci+Kinob} \cdot Z_{NOB} \cdot pHinhibition$
3	Aerobic growth of OHO on Ci	$HMuCiMax \cdot Thetaboho^{T-20} \cdot \dfrac{Ci}{Ci+Ksioho+\dfrac{Ci^2}{Kioho}} \cdot \dfrac{DO}{SWHetroAirOnOff+DO} \cdot \dfrac{NH_3N}{SWNH3\_limit+NH_3N} \cdot \dfrac{PO_4P}{SWPGro\_limit+PO_4P}$
4	Anoxic growth of OHO on Ci with $NO_3N \to NO_2N$	$HMuCiMax \cdot Thetaboho^{T-20} \cdot \dfrac{Ci}{Ci+Ksioho+\dfrac{Ci^2}{Kioho}} \cdot \dfrac{SWHetroAirOnOff}{SWHetroAirOnOff+DO} \cdot \dfrac{SWNH3\_limit}{SWNH3\_limit+NH_3N} \cdot \dfrac{NO_3\text{-}N}{SWAnoxicOnOff+NO_3\text{-}N} \cdot \dfrac{PO_4P}{SWPGro\_limit+PO_4P} \cdot Z_{BH} \cdot \sigma_{ANX} \cdot pHinhibition$
5	Anoxic growth of OHO on Ci with NO_2N	$HMuCiMax \cdot Thetaboho^{T-20} \cdot \dfrac{Ci}{Ci+Ksioho+\dfrac{Ci^2}{Kioho}} \cdot \dfrac{SWHetroAirOnOff}{SWHetroAirOnOff+DO} \cdot \dfrac{SWNH3\_limit}{SWNH3\_limit+NH_3N} \cdot \dfrac{NO_2\text{-}N}{SWAnxNO2OnOff+NO_2\text{-}N} \cdot \dfrac{PO_4P}{SWPGro\_limit+PO_4P} \cdot Z_{BH} \cdot \sigma_{ANX} \cdot pHinhibition$
6	Anoxic growth of OHO on Ci with $NO_3N \to N_2$	$HMuCiMax \cdot Thetaboho^{T-20} \cdot \dfrac{Ci}{Ci+Ksioho+\dfrac{Ci^2}{Kioho}} \cdot \dfrac{SWHetroAirOnOff}{SWHetroAirOnOff+DO} \cdot \dfrac{SWNH3\_limit}{SWNH3\_limit+NH_3N} \cdot \dfrac{NO_3\text{-}N}{SWAnoxicOnOff+NO_3\text{-}N} \cdot \dfrac{PO_4P}{SWPGro\_limit+PO_4P} \cdot Z_{BH} \cdot \sigma_{ANX} \cdot pHinhibition$
7	Aerobic growth of OHO on COD with Ci inhibition	$HMuMax \cdot Thetaboho^{T-20} \cdot \dfrac{Kioho}{Ci+Kioho} \cdot \dfrac{COD}{K_{sCOD}+COD} \cdot \dfrac{DO}{SWHetroAirOnOff+DO} \cdot \dfrac{NH_3N}{SWNH3\_limit+NH_3N} \cdot \dfrac{PO_4P}{SWPGro\_limit+PO_4P} \cdot \dfrac{Z_{BH}}{(HKsCOD+Sbsa+Sbsp+Sbsc+Ci)} \cdot Sbsc \cdot pHinhibition$
8	Anoxic growth of OHO on COD with Ci inhibition with $NO_3N \to NO_2\text{-}N$	$HMuMax \cdot Thetaboho^{T-20} \cdot \dfrac{Kioho}{Ci+Kioho} \cdot \dfrac{COD}{HKsCOD+COD} \cdot \dfrac{SWHetroAirOnOff}{SWHetroAirOnOff+DO} \cdot \dfrac{SWNH3\_limit}{SWNH3\_limit+NH_3N} \cdot \dfrac{NO_3\text{-}N}{SWAnoxicOnOff+NO_3\text{-}N} \cdot \dfrac{PO_4P}{SWPGro\_limit+PO_4P} \cdot \dfrac{Z_{BH}}{(HKsCOD+Sbsa+Sbsp+Sbsc+Ci)} \cdot Sbsc \cdot \sigma_{ANX} \cdot pHinhibition$

(continued)

Table 1 | continued

No.	Biological process	Reaction rates
9	Anoxic growth of OHO on COD with Ci inhibition with NO₂N	$HMuMax \cdot Thetaboho^{T-20} \cdot \dfrac{Kioho}{Ci+Kioho} \cdot \dfrac{COD}{HKsCOD+COD} \cdot \dfrac{SWHetroAirOnOff}{SWHetroAirOnOff+DO} \cdot \dfrac{SWNH3\_limit}{SWNH3\_limit+NH_3N} \cdot SWAnxNO2OnOff \cdot \dfrac{NO_2\text{-}N}{\ }\cdot \dfrac{PO_4\text{-}P}{SWPGro\_limit+PO_4\text{-}P} \cdot \dfrac{Z_{BH}}{(HKsCOD+Sbsa+Sbsp+Sbsc+Ci)} \cdot Sbsc \cdot \sigma_{ANX} \cdot pHinhibition$
10	Anoxic growth of OHO on COD with Ci inhibition with NO₃N → N₂	$HMuMax \cdot Thetaboho^{T-20} \cdot \dfrac{Kioho}{Ci+Kioho} \cdot \dfrac{COD}{HKsCOD+COD} \cdot \dfrac{SWHetroAirOnOff}{SWHetroAirOnOff+DO} \cdot \dfrac{SWNH3\_limit}{SWNH3\_limit+NH_3N} \cdot SWAnxNO2OnOff \cdot \dfrac{NO_2\text{-}N}{\ }\cdot \dfrac{PO_4\text{-}P}{SWPGro\_limit+PO_4\text{-}P} \cdot \dfrac{Z_{BH}}{(HKsCOD+Sbsa+Sbsp+Sbsc+Ci)} \cdot Sbsc \cdot \sigma_{ANX} \cdot pHinhibition$
11	Adsorption of QAC to biomass	$\dfrac{1000}{TSS} \cdot kads \cdot \left(qmax \cdot ThetaAds^{T-15} \cdot \dfrac{KL \cdot Ci}{1+KL \cdot Ci} - \dfrac{Cads}{TSS}\right)^2$

(from day 33 to 381 of the experiment). Figure 2 illustrates the measured *versus* modelled QAC concentrations through the BNR system at QAC feed concentrations of 5, 10, 15, 45 and 60 mg/L. Note that, in Figure 2, the bars for measured data represent the average QAC concentration, and the error bars represent the standard deviation of the measurements. The concentration of QAC is the bulk liquid value; this is a consequence of the influent QAC mass rate and the adsorption/biodegradation rates of QAC in the process (each of which depend on the biomass inventory or SRT of the system). The amount of QAC adsorption to solids was calibrated based on non-active solid concentrations in the pre-anoxic (R1) reactor (since RAS is discharged to the anoxic (R2) reactor), and the amount of QAC removal in R1. Due to adsorption and anoxic growth in the pre-anoxic (R1) and anoxic (R2) reactors, the QAC concentration was always below 2 mg/L in the aerobic reactor (R3); therefore, no impact on nitrification was observed. The modeled results compare favorably with the measured concentrations at all of the feed concentrations tested. Therefore, the model does a reasonable job of predicting QAC adsorption and aerobic/anoxic biodegradation by OHOs.

The experimental nitrification results reported by Yang *et al.* (2014) were used to test the predicted nitrification performance of the mathematical model. The short-term nitrification assays by Yang *et al.* (2014) show the effect of QAC on the removal rate of 100 mg NH4-N/L at 20 °C. The simulations were conducted with constant QAC concentration throughout the test as was the case in the experiments. The simulation results of the short term nitrification assays are illustrated in Figure 3. The lines in Figure 3 represent the modeled N species results; the points represent the measured N species results reported in Yang *et al.* (2014). In the experiments, QAC was observed to inhibit nitrification at a liquid phase concentration of 2 mg/L, as indicated by (1) a low ammonia removal rate and (2) no nitrite accumulation (which indicates a lower than typical AOB growth rate (Dold *et al.* 2015). Figure 3(a) (top chart) shows the N species results when a QAC concentration of 2 mg/L was applied. With a QAC concentration of 2 mg/L, most of the ammonia is oxidized in 48 h and no nitrite accumulation is observed. This low ammonia removal rate and lack of nitrite accumulation indicates a degree of AOB inhibition. Figure 3(b) (bottom chart) shows the N species results when a QAC concentration of 10 mg/L was applied. Ammonia only is reduced from 95 mg/L to 62 mg/L in 96 h. This very low ammonia removal rate and lack of nitrite accumulation indicates

Table 2 | Summary of process stoichiometry

Process	Zbh	Zaob	Znob	Sbsc	NH3N	NO2N	NO3-N	N2	PO4-P	SCO2	Cads	Ci	DO
1		1			$-$fnaob $-$ 1/Yaob	1/Yaob			$-$fpaob	$-1/32$			$-$(3.43-Yaob)/Yaob
2			1		$-$fnnob	$-$1/Ynob	1/Ynob		$-$fpnob	$-1/33$			$-$(1.14-Ynob)/Ynob
3	1				$-$HFzbn				$-$HFzbp	(1-Yh_Ci_Aer)/(Yh_Ci_Aer* MWOxygen2)		-1/Yh_Ci_Aer	$-$(1-Yh_Ci_Aer)/Yh_Ci_Aer
4	1				$-$HFzbn	(1-Yh_Ci Anox)/(gOD_NO3toNO2* Yh_Ci_Anox)	$-$(1-Yh_Ci Anox)/(gOD_NO3toNO2* Yh_Ci_Anox)		$-$HFzbp	(1-Yh_Ci_Anox)/(Yh_Ci_Anox* MWOxygen2)		-1/Yh_Ci_Anox	
5	1				$-$HFzbn	$-$((1-Yh_Ci_Anox)/(gOD_NO2toN2* Yh_Ci_Anox))		(1-Yh_Ci_Anox)/(gOD_NO2toN2* Yh_Ci_Anox)	$-$HFzbp	(1-Yh_Ci_Anox)/(Yh_Ci_Anox* MWOxygen2)		-1/Yh_Ci_Anox	
6	1				$-$HFzbn		$-$(1-Yh_Ci_Anox)/(gOD_NO3toN2* Yh_Ci_Anox)	(1-Yh_Ci_Anox)/(gOD_NO3toN2* Yh_Ci_Anox)	$-$HFzbp	(1-Yh_Ci_Anox)/(Yh_Ci_Anox* MWOxygen2)		-1/Yh_Ci_Anox	
7	1			-1/Yh_Sbsa_Aer	$-$HFzbn				$-$HFzbp	(1-Yh_Cs_Aer)/(Yh_Cs_Aer* MWOxygen2)			$-$(1-Yh_Cs_Aer)/Yh_Cs_Aer
8	1			-1/Yh_Sbsa_Anox	$-$HFzbn	(1-Yh_Cs_Anox)/(gOD_NO3toNO2* Yh_Cs_Anox)	$-$(1-Yh_Cs_Anox)/(gOD_NO3toNO2* Yh_Cs_Anox)		$-$HFzbp	(1-Yh_Cs_Anox)/(Yh_Cs_Anox* MWOxygen2)			
9	1			-1/Yh_Sbsa_Anox	$-$HFzbn	$-$(1-Yh_Cs_Anox)/(gOD_NO2toN2* Yh_Cs_Anox)		(1-Yh_Cs_Anox)/(gOD_NO2toN2* Yh_Cs_Anox)	$-$HFzbp	(1-Yh_Cs_Anox)/(Yh_Cs_Anox* MWOxygen2)			
10	1			-1/Yh_Sbsa_Anox	$-$HFzbn		$-$(1-Yh_Cs_Anox)/(gOD_NO3toN2* Yh_Cs_Anox)	(1-Yh_Cs_Anox)/(gOD_NO3toN2* Yh_Cs_Anox)	$-$HFzbp	(1-Yh_Cs_Anox)/(Yh_Cs_Anox* MWOxygen2)			
11											1	-1	

Table 3 | Summary of rate and stoichiometric constants

Parameter	Value	Parameter	Value
Rate constants			
$MuMax_{AOB}$	0.9	$MuMax_{NOB}$	0.7
$ThetaMu_{AOB}$	1.072	$ThetaMu_{NOB}$	1.06
K_{DOAOB}	0.25	K_{DONOB}	0.5
K_{AOB}	0.7	K_{NOB}	0.1
K_{iAOB}	1.39	K_{iNOB}	200
HMuCiMax	0.26	SWNH3_limit	0.005
$Thetab_{OHO}$	1.029	SWPGro_limit	0.001
Ksi_{OHO}	10	SWHeteroAirOnOff	0.05
Ki_{OHO}	20	SWAnoxicOnOff	0.1
NO3_Ratio	0.4	SWAnoxNO2OnOff	0.01
NO2_Ratio	0.6	Ks_{CO2}	0.1
HKsCOD	5	ThetaAds	0.9884
HMuMax	3.2	KL	0.047
Kads	5	qmax	0.36
Stoichiometric constants			
Y_{AOB}	0.15	Y_{NOB}	0.09
fn_{AOB}	0.07	fn_{NOB}	0.07
fp_{AOB}	0.022	fP_{NOB}	0.022
HFzbp	0.022	Yh_Cs_Anox	0.53
HFzbn	0.07	Yh_Cs_Aer	0.666
Yh_Ci_anox	0.53	Yh_Sbsa_Aer	0.67
Yh_Ci_aer	0.666	Yh_Sbsa_Anox	0.53
gOD_NO3toNO2	1.142	gOD_NO3toN2	2.8556
gOD_NO2toN2	1.713	MWOxygen2	31.998

more severe degree of AOB inhibition. Based on these results, the model accurately predicts the inhibitory threshold of 2 mg/L for AOB conversion of ammonia to nitrite. It is worth noting that the authors' literature search did not reveal data suitable for accurately quantifying NOB inhibition kinetics. However, the model has been structured in a way to allow for NOB inhibition if required/desired.

Table 4 | Comparison of measured versus modeled performance parameters

Parameter	R_3		Effluent	
	Measured	Modeled	Measured	Modeled
pH	7.0 ± 0.4	7.1	7.0 ± 0.4	7.1
TSS (mg/L)	$1{,}309 \pm 124$	1,390	52 ± 2	42
VSS (mg/L)	$1{,}073 \pm 80$	1,180	42 ± 2	37
Soluble COD (mg/L)	273 ± 74	268	282 ± 15	268
NH_3 (mg N/L)	0.9 ± 0.2	0.7	1 ± 0.5	0.7
NO_3 (mg N/L)	23 ± 5	25.5	20 ± 7	25.5
NO_2 (mg N/L)	1.9 ± 0.9	0.2	1 ± 4	0.2

CONCLUSION

A mechanistic mathematical model to describe the fate of QACs in activated sludge processes was developed as an add-on model in BioWin 5.3. The model includes removal of QACs via both biodegradation and adsorption, and quantifies the degree of QAC inhibition on the growth of AOBs, NOBs and OHOs. The model assumes that microorganisms are already acclimated to QACs; no population shifts or changes in metabolic abilities are modeled with time. Simulation results were compared to the studies of Hajaya (2011) and Yang *et al.* (2014) as a means of calibration. The model

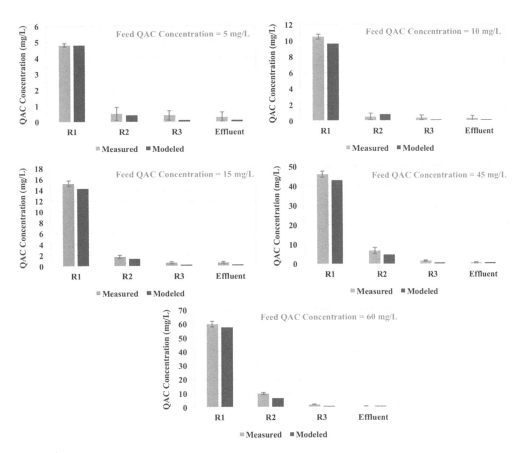

Figure 2 | Measured versus modeled QAC concentrations through the BNR system in Hajaya's study at QAC feed concentrations of 5, 10, 15, 45 and 60 mg/L.

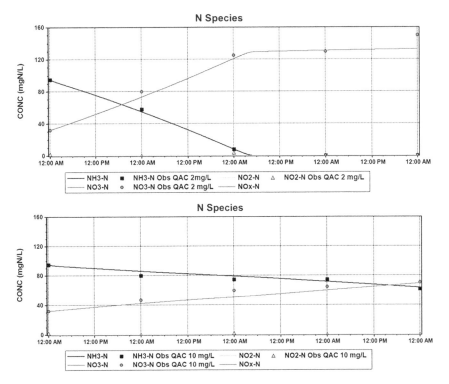

Figure 3 | Measured versus modeled N species responses for Yang *et al.* (2014) short-term nitrification bioassays at QAC batch concentrations of 2 and 10 mg/L.

was found to accurately predict the inhibition of nitrification in batch bioassays and the concentration of QAC in the bulk liquid of a laboratory-scale BNR activated sludge system.

The model provides a preliminary framework for simulating the potential impacts of inhibitory substances on BNR WWTP performance. The following important processes are taken into account by the model:

- Adsorption of QACs onto the biological sludge mass, thereby reducing the potential for inhibition.
- Aerobic and anoxic growth of heterotrophic organisms on QACs according to Haldane kinetics.
- Aerobic and anoxic growth of heterotrophic organisms on soluble influent COD inhibited by the presence of QACs using a Monod-type rate reduction if QACs are present.
- Inhibition of ammonia and nitrite oxidizing organisms using a Monod-type rate reduction if QACs are present.

The model presented here should be viewed as a first step in developing a more comprehensive biological model incorporating inhibition kinetics. Aspects for future investigation could include the following:

- Refinement of the split between adsorption and biodegradation of QACs as removal pathways.
- Refinement of the model to include acclimation of heterotrophic biomass to QACs (e.g. incorporating two populations of heterotrophic biomass [one population that grows slowly on both other COD and QAC in the presence of QAC, and a second specialist population that has a relatively higher growth rate on QAC but can *only* grow on QAC]; using a single population of heterotrophic biomass with a growth rate that changes as a function of the amount of time exposed to QACs).
- Impact of QACs on NOB kinetics.
- Impact of QACs on organisms responsible for biological excess phosphorus removal.
- Impact of temperature on inhibition coefficients.

REFERENCES

Aiba, S., Shoda, M. & Nagatani, M. 1968 Kinetics of product inhibition in alcohol fermentation. *Biotechnology & Bioengineering* 10 (6), 845–864.

Carbajo, J. B., Petre, A. L., Rosal, R., Berna, A., Leton, P., Garcia-Calvo, E. & Perdigon-Melon, J. A. 2015 Ozonation as pre-treatment of activated sludge process of a wastewater containing benzalkonium chloride and NiO nanoparticles. *Chemical Engineering Journal* 283, 740–749.

Chen, M., Zhang, X., Wang, Z., Liu, M., Wang, L. & Wu, Z. 2018 Impacts of quaternary ammonium compounds on membrane bioreactor performance: acute and chronic responses of microorganisms. *Water Research* 134, 153–161.

Dold, P. L., Du, W., Burger, G. & Jimenez, J. 2015 Is Nitrite-Shunt Happening in the System? Are NOB Repressed? In: *WEFTEC Conference Proceedings*, Chicago, Illinois.

Hajaya, M. 2011 *Fate and Effect of Quaternary Ammonium Antimicrobial Compounds on Biological Nitrogen Removal within High-Strength Wastewater Treatment Systems.* PhD Thesis, Georgia Institute of Technology, Atlanta, GA, USA.

Hajaya, M. G. & Pavlostathis, S. G. 2012 Modeling the fate and effect of benzalkonium chlorides in a continuous-flow biological nitrogen removal system treating poultry processing wastewater. *Bioresource Technology* 130, 278–287.

Khan, A. H., Topp, E., Scott, A., Sumarah, M., Macfie, S. M. & Ray, M. B. 2015 Biodegradation of benzalkonium chlorides singly and in mixtures by a *Pseudomonas* sp. isolated from returned activated sludge. *Journal of Hazardous Materials* 299, 595–602.

Ren, R. L., Liu, D., Li, K., Sun, J. & Zhang, C. 2011 Adsorption of quaternary ammonium compounds onto activated sludge. *Journal of Water Resource and Protection* 3 (2), 105–113.

Ruan, T. S., Song, S., Wang, T., Liu, R., Lin, Y. & Jiang, G. 2014 Identification and composition of emerging quaternary ammonium compounds in municipal sewage sludge in China. *Environmental Science & Technology* 48 (8), 4289–4297.

Sütterlin, H. A., Alexy, R., Coker, A. & Kümmerer, K. 2008 Mixtures of quaternary ammonium compounds and anionic organic compounds in the aquatic environment: elimination and biodegradability in the closed bottle test monitored by LC–MS/MS. *Chemosphere* 72 (3), 479–484.

Tezel, U. 2009 *Fate and Effect of Quaternary Ammonium Compounds in Biological Systems.* PhD Thesis, Georgia Institute of Technology, Atlanta, GA, USA.

Yang, J., Tezel, U., Li, K. & Pavlostathis, S. G. 2014 Prolonged exposure of mixed aerobic cultures to low temperature and benzalkonium chloride affect the rate and extent of nitrification. *Bioresource Technology* 179, 193–201.

Zhang, C., Tezel, U., Li, K., Liu, D., Ren, R., Du, J. & Pavlostathis, S. G. 2010 Evaluation and modeling of benzalkonium chloride inhibition and biodegradation in activated sludge. *Water Research* 45 (3), 1238–1246.

Zhang, C., Cui, F., Zeng, G. M., Jiang, M., Yang, Z. Z., Yu, Z. G., Zhu, M. Y. & Shen, L. Q. 2015 Quaternary ammonium compounds (QACs): a review on occurrence, fate and toxicity in the environment. *Science of The Total Environment* 518–519, 352–362.

First received 27 June 2018; accepted in revised form 16 October 2018. Available online 24 October 2018

Towards model predictive control: online predictions of ammonium and nitrate removal by using a stochastic ASM

Peter Alexander Stentoft, Thomas Munk-Nielsen, Luca Vezzaro,
Henrik Madsen, Peter Steen Mikkelsen and Jan Kloppenborg Møller

ABSTRACT

Online model predictive control (MPC) of water resource recovery facilities (WRRFs) requires simple and fast models to improve the operation of energy-demanding processes, such as aeration for nitrogen removal. Selected elements of the activated sludge model number 1 modelling framework for ammonium and nitrate removal were included in discretely observed stochastic differential equations in which online data are assimilated to update the model states. This allows us to produce model-based predictions including uncertainty in real time while it also reduces the number of parameters compared to many detailed models. It introduces only a small residual error when used to predict ammonium and nitrate concentrations in a small recirculating WRRF facility. The error when predicting 2 min ahead corresponds to the uncertainty from the sensors. When predicting 24 hours ahead the mean relative residual error increases to ~10% and ~20% for ammonium and nitrate concentrations respectively. Consequently this is considered a first step towards stochastic MPC of the aeration process. Ultimately this can reduce electricity demand and cost for water resource recovery, allowing the prioritization of aeration during periods of cheaper electricity.

Key words | activated sludge process (ASP), grey-box model, MPC, prediction, stochastic differential equations

Peter Alexander Stentoft (corresponding author)
Thomas Munk-Nielsen
Luca Vezzaro
Krüger A/S, Veolia Water Technologies,
Søborg,
Denmark
E-mail: pas@kruger.dk

Peter Alexander Stentoft
Henrik Madsen
Jan Kloppenborg Møller
Department of Applied Mathematics and
  Computer Science,
Technical University of Denmark,
Kongens Lyngby,
Denmark

Luca Vezzaro
Peter Steen Mikkelsen
Department of Environmental Engineering,
Technical University of Denmark,
Kongens Lyngby,
Denmark

INTRODUCTION

Mathematical modelling of water resource recovery facilities (WRRFs) is a widely established discipline for research, plant design, optimization, simulation of process control strategies, etc. For these purposes there are many models to choose between, such as the activated sludge models (ASM), the anaerobic digestion models (ADM), the University of Cape Town (UCT) model or the TU Delft phosphorous removal model (Wentzel et al. 1992; Henze et al. 2000; Batstone et al. 2002; Gernaey et al. 2004; Meijer 2004; Hu et al. 2007). These complex models have differences in focus and as a result, in their structure. Hence choosing a model structure is, as with all modelling tasks, crucial to the outcome of the project. One important thing to include in the choice of a suitable model is the number of states and parameters. On the one hand more states and parameters leads to a more detailed model. However, on the other hand more details introduce more inputs that need to be distinguished and therefore estimated, measured or, if this is not possible, guessed at

(Vanrolleghem et al. 1995). Furthermore, numerically solving large models with many states leads to long simulation times which can be demanding for data-driven optimizations, which need to be run at short time intervals (seconds to minutes). Although not yet used in the online operation of WWRFs, models can also be used to forecast future variables of interest for use in model predictive control (MPC), which means they should be fast and adaptable to online data.

The activated sludge model number 1 (ASM1) (Henze et al. 1987) describes organic matter degradation, nitrification and denitrification in activated sludge bioreactors. The model contains 13 states variables and 19 parameters. One of the most important challenges in using ASM1 is arguably attributing the many stoichiometric and kinetic parameters (Gernaey et al. 2004). The information needed for the characterization of these can come from three sources (Petersen et al. 2002): (1) default values from the literature, (2) full-scale plant data such as those collected by online sensors, and (3) information obtained from laboratory-scale experiments. The type of data

doi: 10.2166/wst.2018.527

and calibration framework to use is highly dependent on the intended use (e.g. Petersen *et al.* 2002). While (1) might be good for educational purposes or comparison of control strategies (e.g. Gernaey *et al.* 2014), optimization of processes with respect to a specific plant requires (2) and/or (3) (Petersen *et al.* 2002).

MPC aims to predict processes as a function of potential control actions and then choose the best control scenario based on the optimization of some objective function. In WRRFs this can translate to real-time modelling and forecasting of plant performance based on aeration control, optimizing electricity costs and effluent. When it comes to the selection of a suitable model for WRRF MPC strategies, the structure of states and parameters becomes particularly important. This is for the following two reasons: firstly, because parameters should be statistically identifiable from online data to take proper advantage of the real-time setting and secondly, because the computational requirements should be sufficiently low to allow for real-time, recursive simulation of several control scenarios. This means that a good online model should not have strong correlations between parameters, which is the case for parameters of ASM1 (Sharifi *et al.* 2014). Furthermore, the calibration should only depend on online data to avoid delays in updating the model, and hence the recalibration routine should not depend on information obtained from laboratory-scale experiments.

ASM1 has already been simplified to a linearized version to provide faster and yet reliable predictions (Smets *et al.* 2003). Furthermore, ASM1 has been reduced in an effort to make more parsimonious models (e.g. Mulas *et al.* 2007; Cadet 2014). However, the focus of these models is not online operations, i.e. to be updated with online data only. Online model applications are here managed by the use of stochastic, data-driven modelling (DDM) techniques. Many DDM methods exist, depending on the purpose and data availability (e.g. Dürrenmatt & Gujer 2012) and they are generally good alternatives when mechanistic models are not available or not valid (Gernaey *et al.* 2004). Since the detailed mechanistic understanding of the activated sludge process (ASP) already exists, the use of DDMs would ignore all the existing empirical process knowledge of nonlinearities and correlations.

Discretely observed stochastic differential equation (SDE) based models are often referred to as stochastic 'grey-box' (GB) models because the structure of the models represents both the physical/chemical/biological, deterministic ('white-box') understanding of the processes and the statistical, stochastic ('black-box') information

indicated by data. Parameter calibration can be managed in the SDE-GB model by e.g. combining extended Kalman filtering (EKF) techniques and maximum likelihood estimation. This can be done statistically directly from online data by using e.g. the frameworks suggested by Kristensen *et al.* (2004), Tullekin (1993) or Jazwinski (1970). Furthermore, the EKF allows for optimal state estimation and handles additive noise effectively. Thordarson *et al.* (2012), Del Giudice *et al.* (2015) and Carstensen (1994) concluded that in terms of process prediction and control, SDE-GB models of the wastewater processes perform significantly better than traditional black-box models, such as ARMAX models, and also used them to statistically identify Monod-kinetic parameters from online measurements in an ASP (Carstensen *et al.* 1995). SDE-GB models have also been used to model incoming ammonium loads and first-flush phenomena (Bechmann *et al.* 2000; Halvgaard *et al.* 2017), and to forecast rainfall-runoff flow and volume in sewer systems for use in real-time optimization (Thordarson *et al.* 2012; Löwe *et al.* 2016).

In this paper we show that ASM1 can be rewritten as a simpler SDE-GB model that is applicable to online MPC purposes by treating state variables that show only slow and minor changes over short time horizons as model parameters that are kept fixed or intermittently re-estimated using online data. Thus, changes that occur slowly over weeks or months, such as changes in biomass, temperature, maintenance, wastewater composition etc., will be included in the parameters that are re-estimated intermittently with data from the previous few days. The small error introduced by this simplification is estimated by a stochastic diffusion process and consequently it can be managed in the control setup. Following this methodology it is possible to create a stochastic ASM with only three states representing ammonium concentration, nitrate concentration and available oxygen. This model can then be used to optimize the ammonium and nitrate removal process within an MPC approach.

This article presents a simple ASM based on SDEs, which uses flow and aeration data as input and assimilates online measurements of ammonium and nitrate to update model states and thus prepare for providing the best possible forecasts at each time step. The model gives reliable online forecasts of the ammonium and nitrate removal process from a few minutes to up to 24 hours ahead and considers measurement errors. It was developed and tested with data from a small recirculation WRRF with alternating operation. The simplicity of the model makes it a general tool that can be useful in recirculation facilities with different configurations without changing the model setup.

CASE STUDY: NØRRE SNEDE WRRF

The model is developed and tested with data from Nørre Snede WRRF, which is located in central Jutland, Denmark. The plant is designed to handle a maximum capacity of 9,700 population equivalent (PE) and the current load is approximately 4,000 PE.

Operation and design

The WRRF includes several typical treatment processes that the wastewater goes through before discharge. Listed in order from when the wastewater enters the process, these are pretreatment, grit removal and grease trap, chemical dosage, nitrification/denitrification and secondary treatment. The nitrification/denitrification in the Nørre Snede plant happens in a process tank with a total volume of 3,500 m³. The tank is divided into three smaller chambers operating under different conditions. This is illustrated in Figure 1, which also shows that the aeration tank is equipped with nutrient sensors, aeration equipment, a recirculation pump and rotors that control the flow direction/velocity (direction shown with arrows in the figure; rotors are located at the bridge).

The facility is currently operated with a rule-based control strategy, as described by Isaacs & Thornberg (1998), Zhao *et al.* (2004) and Kim *et al.* (2014), for example. In this case the control switches aeration on/off as a function of online ammonium and nitrate concentration measurements. Therefore the conditions switch between anoxic and aerobic and the cycle lengths depend on the conditions

in the process tanks. This is managed in the control platform STAR Utility Solutions™ (Sørensen *et al.* 1994; Nielsen & Önnerth 1995).

Data

The current control of the plant (i.e. actuator settings controlling aeration and inlet flow) is updated every 2 min and as result, aeration and inflow data are available every 2 min. The control rules are based on ammonium and nitrate signals, which are only sampled every 5 min directly in the aeration tank, meaning that observations are sampled at a different frequency to the control sampling. Calibration of sensors happens automatically two to four times per day, resulting in 30–60 min with no new observations. There is a response time from when aeration starts/stops until this is observed by the sensors. This is due to hydraulics in the tank and processing time in the sensors (Rieger *et al.* 2003). This response time is estimated using the method suggested by Stentoft *et al.* (2017): an estimate from the point at which conditions are shifting to the point at which a change in the trend in measurements is observed. Flow data are available at the outlet (after the settler) and change between 0 and ~45 m³/h because of a pumping scheme. To account for this scheme, flow data are filtered by a second-order Fourier series. The available data used in this study are summarized in Table 1.

THEORY AND METHODOLOGY

This section includes a description of the SDE-GB model, a simplification of the ASM1 model with noise terms added, the inclusion of aeration control and inflow as input. Finally, the parameter estimation is briefly described.

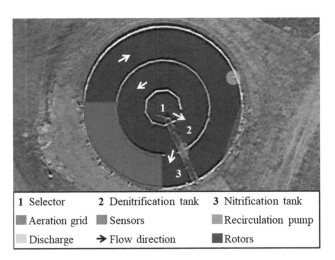

1 Selector	2 Denitrification tank	3 Nitrification tank
▦ Aeration grid	▦ Sensors	▦ Recirculation pump
▦ Discharge	➜ Flow direction	▦ Rotors

Figure 1 | Overview of the process tank at Nørre Snede WRRF with important features labelled.

Table 1 | Overview of online data used in this study

Symbol	Description	Sample frequency	Unit	Uncertainty
Q	Effluent Flow	2 min	m³/L	Unknown
O	Actuator setting	2 min	mgO₂/L	0
MsNH	Measured ammonium	5 min	mgN/L	±3%
MsNO	Measured nitrate	5 min	mgN/L	±5%

The uncertainty is based on the information available from the sensor manuals (HACH Lange Aps 2013, 2014).

Stochastic grey-box model

A discretely observed stochastic grey-box model can be written on the state-space form as

$$dx_t = f(x_t, u_t, t, \theta)dt + \sigma(u_t, t, \theta)d\omega$$

$$y_k = h(x_k, u_k, t_k, \theta) + e_k$$

where the description of the dynamics of the states x_t are divided into a (deterministic) drift term $f(x_t, u_t, t, \theta)dt$ and a (stochastic) diffusion term $\sigma(u_t, t, \theta)d\omega$. The system is observed only through y_k which is linked to x_t via the observation equation $h(x_k, u_k, t_k, \theta)$. The residual error is separated in to two terms. *Diffusion*, $\sigma(u_t, t, \theta)d\omega$ represents model approximations and undescribed noise disturbances, i.e. changes in biomass efficiency, unmodelled inflow, uncertainty of input variables (u_t), or true stochastic behaviour in the processes. *Measurement noise*, e_k represents the noise occurring due to imperfect accuracy of the measuring equipment, i.e. measurement uncertainty in the ammonium and nitrate sensors.

Simplification of ASM1

Following the notation proposed by Corominas *et al.* (2010) the ordinary differential equations that govern ammonium, nitrate and oxygen in ASM1 can be written in a Gujer matrix, as presented in the Supplement (available with the online version of this paper), together with a brief description of ASM1 parameters and state variables. The complexity of these equations is considered an obstacle for use in a real-time setting since many of the variables are unmeasured and consequently constants would be difficult to distinguish. We therefore make simplifications to obtain a more suitable model. The main assumption in this simplification is that the model parameters will be re-estimated frequently, and therefore several state variables of ASM1 will become constant and some parameters will become unimportant.

- The rate, ρ_4, which governs ammonification of soluble organic nitrogen can be ignored. This is considered reasonable as the ammonification rate parameter k_{am} is typically small compared to the process rates of nitrification and denitrification (Henze *et al.* 1987).
- The state variable S_B is constant throughout the day. In practice it will follow a diurnal pattern similar to that of S_{NH4}, but since S_B is not measured these will be difficult to distinguish.
- The state variables governing active heterotrophic and autotrophic biomasses X_{OHO} and X_{ANO} can be considered

constant on a daily basis, and hence can be treated as parameters. This is considered reasonable as the biomass is known to only change over longer periods of time.
- The parameter for the relative amounts of N/COD in biomass, i_{N_COD}, can be ignored. A stoichiometric calculation by Henze *et al.* (1987) (assuming a typical cell formation, $C_5H_7O_2N$) indicates that i_{N_COD} is 0.086. This is very small compared to $1/Y_{ANO}$ which is approximately 4.2, and therefore it will be difficult to estimate.
- The half-velocity parameters $K_{O2,OHO}$ and $K_{O2,ANO}$ for oxygen use are considered to be equal ($K_{O2,OHO} = K_{O2,ANO} = K_{O2}$), as Henze *et al.* (1987) argue that they are not, quantitatively, that different.

Applying these assumptions, the new, simple model of the ASP can be identified. The shorthand notations α_{NH4}, α_{NO3} and α_{O2} refer to the changes in concentration of ammonium, nitrate and relative oxygen respectively.

$$S'_{NH_4} \approx \alpha_{NH_4} = -\theta_1 \left(\frac{S_{NH_4}}{K_{NH_4,ANO} + S_{NH_4}} \right) S_{O,MO}$$

$$S'_{NO_3} \approx \alpha_{NO_3} = \theta_1 \left(\frac{S_{NH_4}}{K_{NH_4,ANO} + S_{NH_4}} \right) S_{O,MO}$$
$$- \theta_2 \left(\frac{S_{NO_3}}{K_{NO_3,OHO} + S_{NO_3}} \right) (1 - S_{O,MO})$$

$$S'_{O,MO} \approx \alpha_{O_2} = -\left(\theta_3 + \theta_4 \left(\frac{S_{NH_4}}{K_{CNH_4,ANO} + S_{NH_4}} \right) \right) S_{O,MO}$$

The half-saturation constant $Kc_{NH_4,ANO}$ is introduced because the state $S_{O,MO}$ is the Monod term indicating how quickly the process is running relative to the maximum rate

$$S_{O,MO} = \left(\frac{S_{O_2}}{K_O + S_{O_2}} \right)$$

The seven new parameters to estimate online are therefore θ_i, $K_{NO3,OHO}$, $K_{NH4,ANO}$ and $Kc_{NH4,ANO}$ where θ_i relates to the original ASM1 parameters as

$$\theta_1 = \frac{1}{Y_{ANO}} \mu_{ANO,Max} X_{ANO}$$

$$\theta_2 = \frac{1 - Y_{OHO}}{2.86 Y_{OHO}} \mu_{OHO,Max} \left(\frac{S_B}{K_{S_B} + S_B} \right) \eta_{\mu OHO,Ax} X_{OHO}$$

$$\theta_3 = \frac{1 - Y_{OHO}}{Y_{OHO}} \mu_{OHO,Max} \left(\frac{S_B}{K_{S_B} + S_B} \right) X_{OHO} C_1$$

$$\theta_4 = \frac{4.57 - Y_{ANO}}{Y_{ANO}} \mu_{ANO,Max} X_{ANO} C_2$$

where C_1 and C_2 are correction factors that are introduced because $S_{MO,O}$ is the relative amount of oxygen.

Aeration control and inflow

For the purpose of using the model for MPC of N-removal, it is necessary to include the effect of aeration and incoming wastewater as external inputs in the model. The signal determining the intensity of aeration and measurements of incoming wastewater flow are available online and consequently the control should be a function of these, i.e. $C\_(O,Q)$ is a function that describes the effect of inflow and aeration control on the given state. These functions are here determined from the literature. More specifically, the two-films theory (Lewis & Whitman 1924) and diurnal variations in ammonium concentration and constant (low) nitrate concentrations in the incoming wastewater (Henze & Comeau 2008). This means that $C_{NH4}(O,Q)$ and $C_{NO3}(O,Q)$ are given as

$$C_{NH_4}(O, Q) = (r_c + \rho Q)(\mu_{NH_4,in} + \Sigma_{i=1}^{n=2}[s_i \sin(iwt)$$
$$+ c_i \cos(iwt)] - S_{NH_4})C_{NO_3}(O, Q) = (r_c + \rho Q)(\mu_{NO_3,in} - S_{NO_3})$$

where $\mu_{NH4,in}$, $\mu_{NO3,in}$, s_i and c_i are parameters relating to the inflow. Note that $\mu_{NO3,in}$ is typically ~0 (e.g. Henze & Comeau 2008). The parameters r_c and ρ are related to the recirculation to and the volume of the aeration tank (see Figure 1). The control of the aeration is given as

$$C_{O,MO}(O, Q) = k_1 O(S_{O,MOmax} - S_{O,MO})$$

where k_1 is a transfer constant (Lewis & Whitman 1924) related to the size and efficiency of the aeration equipment. The maximum value of the relative oxygen state is 1, and hence $S_{O,MOmax}$ is set to 1 and should not be estimated.

Stochastic ASM

A three-state grey-box model governing the ammonium and nitrate concentrations in the aeration tank can be written as

$$dS_{NH_4} = f_{NH4}(..)dt + \sigma(u_t, t, \theta)d\omega_1$$
$$= \alpha_{NH_4}dt + C_{NH_4}(O_t, Q_t)dt + \sigma_{11}d\omega_1$$

$$dS_{NO_3} = f_{NO3}(..)dt + \sigma(u_t, t, \theta)d\omega_2$$
$$= \alpha_{NO_3}dt + C_{NO_3}(O_t, Q_t)dt + \sigma_{22}d\omega_2$$

$$dS_{O,MO} = f_{O,MO}(..)dt + \sigma(u_t, t, \theta)d\omega_3$$
$$= \alpha_{O_2}dt + C_{O,MO}(O_t, Q_t)dt + \sigma_{33}d\omega_3$$

where the deterministic terms α and C govern the ASP, the aeration and the inflow as described in previous sections. To avoid negative noise and to make estimation of small noise processes easier, the diffusion terms are estimated as exponential parameters (i.e. $\sigma_{ii} = exp(s_{ii})$, $i \in [1, 2, 3]$). The system is discretely observed through ammonium and nitrate sensors in the aeration tank. The measurements (*MsNH* and *MsNO*) from these relate to the system as

$$MsNH = S_{NH_4} + exp(s_{1,NH4})\epsilon_{NH4}$$
$$MsNO = S_{NO_3} + exp(s_{1,NO3})\epsilon_{NO3}$$

where ϵ_{NH4} and ϵ_{NO3} are independent and identically distributed (i.i.d.) $N(0, 1)$, i.e. the residuals of the measurements are normally distributed with zero mean and $exp(s_{1,NH4})$, $exp(s_{2,NO3})$ standard deviation. The changes in the states dS_{NH4}, dS_{NO3} and $dS_{MO,O}$ are given as state variables where ω_1, ω_2 and ω_3 are 1-dimensional standard Wiener processes and $exp(s_{11})$, $exp(s_{22})$ and $exp(s_{33})$ represent the deviation of these processes.

Online parameter estimation in stochastic ASM

The stochastic model presented fits the general model structure for continuous-discrete stochastic state space models, i.e. a model of the state variables in continuous time and measurements of some of the states at discrete times. The R-package CTSM-R (Kristensen *et al.* 2004; Juhl *et al.* 2016) can manage just this kind of system, and is therefore used to estimate parameters and predict the effect of control. This paper provides only a brief summary of how the package works and how it is used here. For further information on this, see CTSM-R (2018).

The parameter estimates are based on a maximum likelihood method, by assuming Gaussian-distributed conditional probability densities.

$$L(\theta; Y_N) = \prod_{t=1}^{N} \frac{\exp(-0.5\epsilon_t^T R_{t|t-1}^{-1}\epsilon_t)}{\sqrt{\det(R_{t|t-1})2\pi}} p(y_0|\theta)$$

where $\epsilon_t = y_t - \hat{y}_{t|t-1}$ ($\hat{y}_{t|t-1} = E(y_t|y_{t-1}, \theta)$) and $R_{t|t-1} = V(y_t|y_{t-1}, \theta)$. ϵ_t and $R_{t|t-1}$ are computed by means of a version of the EKF (e.g. Jazwinski 1970). The likelihood function can be simplified to a simpler log-likelihood function by conditioning on y_0 and taking the negative logarithm. However, this rewriting is omitted here. The parameter estimates are then obtained by minimizing this log-likelihood.

The EKF mentioned above is a continuous-discrete time version of the EKF. With initial conditions for the model values and variance estimate ($\hat{x}_{1/0}$ and $P_{1/0}$) the filter approximations of the output predictions are given as

$$\hat{y}_{k|k-1} = h(\hat{x}_{k|k-1}, u_k, t_k, \theta) \quad R_{k|k-1} = CP_{k|k-1}C^T + S_k$$

which here translates to

$$\hat{y}_{k|k-1} = \begin{cases} \hat{s}_{\text{NH}_4 k|k-1} \\ \hat{s}_{\text{NO}_3 k|k-1} \end{cases} \quad R_{k|k-1} = \begin{bmatrix} 1 & 0 & 0 \\ 0 & 1 & 0 \end{bmatrix} \begin{bmatrix} p_{11} & p_{12} & p_{13} \\ p_{21} & p_{22} & p_{23} \\ p_{31} & p_{32} & p_{33} \end{bmatrix}$$

$$\begin{bmatrix} 1 & 0 \\ 0 & 1 \\ 0 & 0 \end{bmatrix} + \begin{bmatrix} s_{1,NH_4} & 0 \\ 0 & s_{1,NO_3} \end{bmatrix} = \begin{bmatrix} p_{11} + s_{1,NH_4} & p_{12} \\ p_{21} & p_{22} + s_{1,NO_3} \end{bmatrix}$$

Here C is the Jacobian of the observation equation, h. The innovation given by

$$\epsilon_k = y_k - \hat{y}_{k|k-1}$$

The Kalman gain, K_k, for the EKF is then calculated as

$$K_k = P_{k|k-1}C_k^T(R_{k|k-1})^{-1}$$

and the system is updated

$$\hat{x}_{k|k} = \hat{x}_{k|k-1} + K_k\epsilon_k$$

$$P_{k|k} = P_{k|k-1} - K_k R_{k|k-1} K_k^T$$

$\hat{x}$ is the state estimates (S_{NH4}, S_{NO3} and $S_{O,MO}$). The state prediction is done numerically by

$$\frac{d\hat{x}_{t|k}}{dt} = f(\hat{x}_{k|k-1}, u_t, t, \theta), \quad t \in [t_k, t_{k+1}]$$

$$\frac{dP_{t|t_k}}{dt} = A(t)P_{t|t_k} + P_{t|t_k}A(t)^T + \sigma(t)\sigma(t)^T, \quad t \in [t_k, t_{k+1}]$$

where $A(t)$ is the Jacobian of the drift term $f_i(t,...)$. This Jacobian is calculated using a method based on Speelpenning (1980). In calculations of the Jacobian it is assumed that $x = \hat{x}k/k-1$, $u = uk$, $t = tk$ and the parameters, θ, are known. The ordinary differential equations (ODEs) are solved by numerical integration schemes suggested by Hindmarsch (1983) (cited in Kristensen & Madsen 2003, p. 17). This is to ensure an intelligent re-evaluation of A and σ. From this construction we see that the approximation is only good when nonlinearities are not too strong.

The estimation setup implies that initial state and parameter estimates are necessary in the parameter estimation procedure. These can be supplied either as prior distributions or simply as estimates with some maximum and minimum boundaries. For most parameters these initial estimates are based on the literature (e.g. Henze *et al.* 1987; Henze & Comeau 2008). However, a few parameters are unnecessary or cannot be estimated. The initial state values $S_{NH4,0}$, $S_{NO3,0}$ and $S_{O,MO,0}$ can easily be estimated directly from data, as ammonium and nitrate are directly measured and the oxygen signal is known, and hence these are considered unnecessary to estimate. Furthermore, the parameters $K_{NO_3,OHO}$, θ_3 and θ_4 show very small deviations and notably correlation with other parameters. It is also argued by Henze *et al.* (1987) that $K_{NO_3,OHO}$ does not need estimating. For these reasons these are kept constant here at $[K_{NO3,OHO}, \theta_3, \theta_4] = [3.0, 5.0E-6, 1.0]$, thereby reducing the number of parameters to estimate.

RESULTS AND DISCUSSION

Firstly the model is qualitatively evaluated by comparing the model predictions with data and discussing parameter estimates. Secondly, the model is quantitatively evaluated by running it for 1 month and discussing the statistics of the residuals. We stress that the model is run 'online' in the sense that parameters are estimated only by minimizing the objective function described in the previous section. Furthermore, the states S_{NH4}, S_{NO3} and $S_{O,MO}$, are updated using the EKF whenever a new measurement becomes available. Figure 2 shows an example of one prediction of ammonium at a given inlet flow and aeration signal. The state, S_{NH4}, is updated with present data and then predicted 2 hours ahead. Clearly, uncertainty increases with increasing forecast horizon.

Model dynamics

Parameters are estimated with data from a period at the beginning of October 2016, chosen arbitrarily among periods without rain. The length of the parameter estimation period is 4 days and 4 hours (corresponding to 3,000 time steps of 2 min). These parameters are used to predict the concentrations of ammonium and nitrate in the aeration tank. Figures 3 and 4 show predictions of the ASP 60 time steps ahead corresponding to 2 hours, given the aeration signal. This is done for 24 hours, meaning that each time a new measurement becomes available a prediction similar to Figure 2 is made and compared with

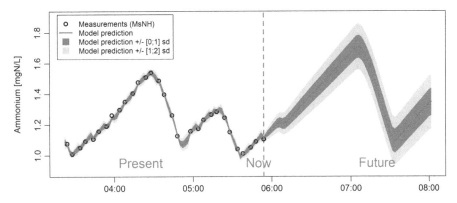

Figure 2 | An example of a 2-hour prediction of ammonium concentrations (which is run every 2 min in the online setup). The uncertainty increases further into the future, as estimated by the SDE.

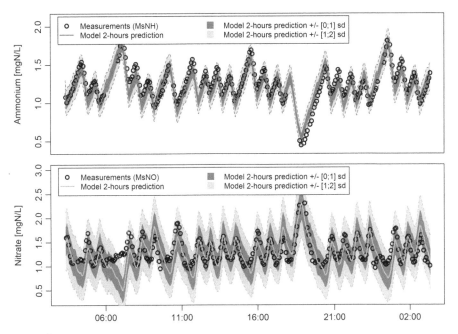

Figure 3 | 2-hour predictions (60 timesteps of 2 mins) of the ammonium concentration in the aeration tank (S_{NH4}) with the measured concentrations (top) and of the nitrate concentration in the aeration tank (S_{NO3}) with the measured concentrations (bottom). Note that the y-axes differ because there is more variation in the nitrate observations.

data. This is done during normal operation of the plant i.e. no rain and no (known) problems.

In Figure 3 it is evident that under normal operation, the modelled ammonium concentration follows the same dynamics as the data. During the aeration phase the ammonium concentration decreases, and when aeration is switched off, NH_4 increases. In periods when no new data are received (i.e. calibration of the ammonium sensor from 17:30 to 18:30), the model continues to provide reliable estimates. The nitrate concentrations estimated in Figure 3 also follow dynamics similar to those in the data. It is noted that when aeration is off, nitrate decreases and when it is on, it

increases. However, a dynamic starting at 06:00 does not follow the behavior shown by the sensor measurements. This period contains a relatively long timespan without aeration that would normally mean denitrification, but in this case we see that nitrate increases. This could be due to some unmodelled dynamics, problems with a drifting nitrate sensor or an unusually large load of nitrate in the influent from industry, for example. Overall, the results show that the uncertainty of the nitrate predictions is greater than the uncertainty of the ammonium predictions, and hence larger deviations from the modelled concentrations are expected. Figure 4 shows the estimates of the unmeasured

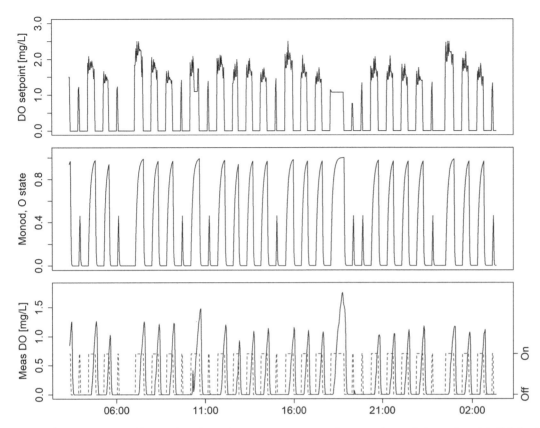

Figure 4 | Top: the input, O. Middle: the estimated Monod oxygen state, $S_{O,MO}$. Bottom: the measured oxygen in the aeration tank (solid line) and the binary signal for aeration on/off (dotted line).

state, $S_{O,MO}$. It is plotted together with the measured dissolved oxygen (DO) concentrations and the setpoint of the actuator, O. It is clear that when the setpoint is lower, it takes a longer time for $S_{O,MO}$ to reach the maximum level. The comparison between the aeration status and the measured oxygen concentrations highlight that short periods of aeration are not registered in the measurements. This is caused by the location of the sensor, which is opposite the aeration grid (see Figure 1). This supports the choice of not including the measured DO as input/state in the model. Therefore the actuator signal is considered superior as it reports all periods of added oxygen and furthermore, it does not have any response time from when air is added until it is observed in the tank.

Parameter estimates

The parameters that are estimated in the period mentioned previously are presented in Table 2.

Some parameters are difficult to evaluate as they represent some catchment/plant-specific information. Nonetheless,

parameters are discussed here and in some cases compared with the literature:

- The residual errors (which are split into diffusion and measurement noise) are found to be similar to what is estimated for ammonium in the aeration tank by Halvgaard *et al.* (2017) in another plant with similar sensors and configuration. However, the uncertainty of the measuring equipment is much smaller compared to what is stated by the sensor manufacturer (see Table 1 footnote).

- The Monod kinetics parameter, $K_{NH4,ANO}$, is found to be similar to what is statistically estimated by Carstensen *et al.* (1995). Here it was found to be between 0.44 and 0.76 among different plants and catchments. However, it is lower than ~1, which is what is suggested by Henze *et al.* (1987).

- The mean incoming concentrations of ammonium, $\mu_{NH4,in}$ and nitrate, $\mu_{NO3,in}$ seem reasonable as these are similar to what is typically found in a 'catchment with little industrial activity' according to Henze & Comeau (2008). The diurnal variations in ammonium, s_1, s_2, c_1

Table 2 | Parameter estimates and standard deviations

Parameter	Estimate	Standard deviation
Simplified ASM		
$K_{NH4,ANO}$ [mgN/L]	4.80×10^{-1}	1.80×10^{-1}
$K_{CNH4,ANO}$ [mgN/L]	1.05×10^{-3}	6.99×10^{-4}
θ_1 [mgN/L]	5.00×10^{-2}	5.39×10^{-3}
θ_2 [mgN/L]	2.47×10^{-1}	1.59×10^{-2}
Aeration control and inflow		
s_1 [mgN/L]	1.14	6.63×10^{-1}
s_2 [mgN/L]	-5.63×10^{-1}	3.73×10^{-1}
c_1 [mgN/L]	4.25×10^{-1}	3.24×10^{-1}
c_2 [mgN/L]	-5.82×10^{-2}	1.75×10^{-1}
$\mu_{NH4,in}$ [mgN/L]	1.35×10^{-1}	6.76
$\mu_{NO3,in}$ [mgN/L]	1.10×10^{-2}	2.01×10^{-2}
K_1 [L/mgO]	1.12×10^{-1}	6.48×10^{-3}
ρ[]	1.08×10^{-5}	1.64×10^{-5}
r_c []	7.21×10^{-4}	4.31×10^{-4}
Noise and diffusion terms		
s_{11} [log(mgN/L)]	-4.79	7.65×10^{-2}
s_{22} [log(mgN/L)]	-3.20	2.06×10^{-2}
s_{33} []	-2.86	2.16×10^{-1}
$s_{1,NH}$ [log(mgN/L)]	-8.66	1.26×10^{-1}
$s_{1,NO}$ [log(mgN/L)]	-1.80×10^{-1}	2.56×10^{-1}

Parameters are calculated using 3,000 timesteps of data corresponding to 4 days and 4 hours. The period is from the beginning of October 2016.

and c_2 are catchment specific. However, these are considered reasonable as they produce a variation that is comparable with the one presented in Henze & Comeau (2008), i.e. a similar shape with peaks in the morning and afternoon.

- The new parameters θ_1 and θ_2 are difficult to compare with the literature as they depend on many parameters and the states X_{ANO} and X_{OHO}. However, following the typical parameter suggestions from Henze *et al.* (1987), these should be 3.33 X_{ANO} and 0.82 X_{OHO} respectively.
- The parameters related to the incoming water, r_c and ρ, are estimated to be 7.21×10^{-4} and 1.08×10^{-5} respectively. Summing these and multiplying the mean flow with ρ we get 1.11×10^{-3}. This is slightly more than the expected value of 3.43×10^{-4} which is found by dividing mean flow with the volume of the process tank. This difference can be due to the recirculation that happens between the nitrification and denitrification tanks.
- The relative oxygen transfer rate k_1 depends on many factors such as tank design (e.g. reactor geometry, aeration design), physico-chemical properties (e.g. liquid composition, viscosity, temperature) and the presence of biomass (e.g. Pittoors *et al.* 2014). Therefore it is difficult to determine empirically as it varies between facilities and over time.

Model performance

The model's predictive ability is tested by re-estimating parameters every hour for a period of 1 month and 1 week, starting late September 2016. The model is then used to predict concentrations of ammonium and nitrate 1, 60 and 720 time steps ahead (corresponding to 2 min, 2 hours and 1 day, respectively). The predictions are compared with data and a 24-hour running mean absolute residual is calculated. Figure 5 illustrates how this changes over time for predictions 2 hours ahead (60 time steps). Table 3 shows the statistics of the mean absolute residual for all the different prediction horizons.

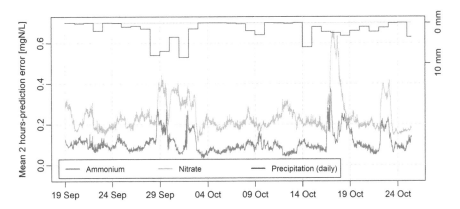

Figure 5 | Mean absolute error for the 2-hour (60 timesteps of 2 mins) prediction of ammonium and nitrate. Plotted with daily precipitation data from DMI (2018).

Table 3 | Summary statistics of the residuals of the model predictions, based on residuals obtained from using the model with data from Nørre Snede WRRF for a period of 1 month in September/October 2016

Precipitation[a]	Prediction –	Mean absolute error [mgN/L]		Relative absolute error [–]	
		No	Yes	No	Yes
Ammonium	2 min	0.0264	0.0278	0.0212	0.0223
	2 hours	0.0884	0.1065	0.0709	0.0855
	24 hours	0.0925	0.129	0.0855	0.104
Nitrate	2 min	0.0804	0.081	0.0594	0.0598
	2 hours	0.212	0.2399	0.157	0.177
	24 hours	0.246	0.374	0.1815	0.276

'Precipitation' indicates whether only dry weather periods (No) or all periods are considered (Yes).
[a]Wet weather periods are as follows: 28 September to 2 October, 15–19 October, 22–23 October.

In Figure 5 it is evident that during some periods (i.e. October 2nd, October 15th and October 22nd) the uncertainty increases. Comparison with rain data supplied by the Danish Meteorological Institute (DMI 2018) shows that many of these periods are characterized by wet weather. This is also indicated in Table 3 where the general picture is that uncertainty increases during wet weather. In Table 3 it is also clear that the relative 2 min uncertainty is ~2% for ammonium and ~6% for nitrate. This is comparable with the sensor uncertainty listed in Table 1 and indicates that the 2 min predictions cannot be further improved even with a more detailed deterministic model. The 2-hour and 24-hour predictions perform worse than the sensor uncertainty. This can, on the one hand, indicate that there is room for improvement, but can, on the other hand, also mean that there is some stochastic behavior (e.g. incoming nutrients or biomass efficiency) which is more pronounced when predicting further ahead from the EKF state update. It should be added that the treatment requirements for Nørre Snede WRRF state that during a 24-hour period, the ammonium concentration should be <2 mgN/L and total N should be <8 mgN/L in the outlet. Consequently this means that accurately predicting ammonium is more important because nitrate only affects total N. The relative uncertainties of ammonium and nitrate of <10% and <20% respectively (in dry weather, 24 hours ahead) are considered sufficient for stochastic MPC.

Towards MPC – simplified and full ASM models

The results in Table 3 are difficult to compare with full ASM models, as to our knowledge there is no framework for making the full ASM models online adaptive to data. However, our results can be compared with data-driven model parameter estimations of full ASMs (i.e. other methods that rely only on data from online ammonium/nitrate sensors). One example of such an approach is provided by Sin et al. (2008), where the parameters of an ASM2d model were estimated using only frequently sampled online ammonium, nitrate and oxygen measurements (sampled every 5 min, similar to this study) in the 50,000 PE Haaren WRRF in the Netherlands with alternating control of aeration. Parameters were calibrated using Monte Carlo simulations to minimize a weighted sum of squared errors (WSSE) based on a calibration period of 16,117 measurements (56 days). Model performance was compared with data in a validation period of 9,217 measurements (32 days). The results showed a mean absolute error (MAE) for ammonium of 1.39 mgN/L and 0.98 mgN/L in the calibration and validation periods, respectively. For nitrate concentrations an MAE of 2.56 mgN/L and 2.31 mgN/L were found. These error values are 10 times larger than what we found in this study, cf. Figure 3. Also, the framework presented in Sin et al. (2008) differs from this study in that the model states are not updated when new data become available and hence short time predictions in the validation period (up to 24 hours ahead) are not based on all the information available in an online situation. Additionally it is noted that the framework by Sin et al. (2008) required some computation time (45 min per simulation (in 2008) on a PC, and 500 Monte Carlo simulations where used to obtain a model), which makes it non-ideal for online applications.

The development of tools for online performance optimization of WRRFs using models is crucial for exploiting the full potential of digitalization. Hence, the development of robust approaches to online identifiable ASMs for improved short horizon predictions is needed. These models should also include additional processes, such as biological removal of COD and P, to achieve an overall improvement of all the removal processes in the plant. This paper provides a first step in this direction with online predictions of ammonium and nitrogen removal.

CONCLUSION

Grey-box models based on SDEs are efficient tools as they can estimate both processes and noise from real-time data. A stochastic model of an aeration tank is proposed here. The model contains a deterministic term consisting of both a simplified ASM and input functions determining the influence of control and inflow. The model is used to predict the nitrification/denitrification in Nørre Snede WRRF in Denmark as a function of aeration and inflow.

The results show that despite the simple structure of the proposed model, the dynamics of the nutrient concentrations are captured. Quantitative investigations show that the processes are predicted accurately, i.e. 24-hour predictions of the ammonium and nitrate concentrations in the aeration tank are predicted with relative errors of $<10\%$ and $<20\%$ respectively. Consequently this is considered to be a step towards stochastic MPC of water resource recovery processes.

ACKNOWLEDGEMENTS

This work was partly funded by the Innovation Fund Denmark (IFD) under File No. 7038-00097B – the first author's industrial PhD: *Stochastic Predictive Control of Wastewater Treatment Processes*. Data for this study were supplied by Ikast Brande Forsyning A/S.

REFERENCES

Batstone, D. J., Keller, J., Angelidaki, R. I., Kalyuzhnyi, S. V., Pavlostathis, S. G., Rozzi, A., Sanders, W. T. M., Siegrist, H. & Vavilin, V. A. 2002 *Anaerobic Digestion Model No. 1. Scientific and Technical Report No. 13*. IWA Publishing, London, UK.

Bechmann, H., Madsen, H., Kjølstad Poulsen, N. & Nielsen, M. K. 2000 Grey box modelling of first flush and incoming wastewater at a wastewater treatment plant. *Environmetrics* **2000** (11), 1–12.

Cadet, C. 2014 Simplifications of Activated Sludge Model with preservation of its dynamic accuracy. *IFAC* **47** (3), 7134–7139.

Carstensen, J. 1994 *Identification of Wastewater Treatment Processes*. PhD Thesis, IMSOR, Technical University of Denmark.

Carstensen, J., Harremoës, P. & Madsen, H. 1995 Statistical identification of monod-kinetic parameters from on-line measurements. *Water Science and Technology* **31** (2), 125–133.

Corominas, L. L., Rieger, L., Takács, I., Ekama, G., Hauduc, H., Vanrolleghem, P. A., Oehmen, A., Gernaey, K. V., van Loosdrecht, M. C. M. & Comeau, Y. 2010 New framework for standardized notation in wastewater treatment modelling. *Water Science and Technology* **61** (4), 841–857.

CTSM-R 2018 *CTSM-R – Continuous Time Stochastic Modelling for R*. http://ctsm.info, visited 31 January 2018.

Del Giudice, D., Lowe, R., Madsen, H., Mikkelsen, P. S. & Rieckermann, J. 2015 Comparison of two stochastic techniques for reliable urban runoff prediction by modeling systematic errors. *Water Resources Research* **51** (7), 5004–5022.

DMI 2018 Vejrarkiv. https://www.dmi.dk/vejr/arkiver/vejrarkiv/, visited 17 January 2018.

Dürrenmatt, D. & Gujer, W. 2012 Data-driven modeling approaches to support wastewater treatment plant operation. *Environmental Modelling and Software* **30**, 47–56. http://dx.doi.org/10.1016/j.envsoft.2011.11.007.

Gernaey, K. V., van Loosdrecht, M. C. M., Henze, M., Lind, M. & Jorgensen, S. B. 2004 Activated sludge wastewater treatment plant modelling and simulation: state of the art. *Environmental Modelling and Software* **19** (9), 763–783.

Gernaey, K. V., Jeppsson, U., Vanrolleghem, P. A. & Copp, J. B. 2014 *Benchmarking of Control Strategies for Wastewater Treatment Plants, Scientific and Technical Report Series*. IWA Publishing Company, London, UK.

HACH Lange APS 2013 *Amtax sc, Amtax indoor sc Brugsanvisning*, 8th edn. HACH Lange APS.

HACH Lange APS 2014 *Nitratax sc – Betjeningsvejledninger*, 6th edn. HACH Lange APS.

Halvgaard, R. F., Vezzaro, L., Grum, M., Munk-Nielsen, T., Tychsen, P. & Madsen, H. 2017 *Stochastic Greybox Modeling for Control of an Alternating Activated Sludge Process*. (DTU Compute-Technical Report-2017, Vol. 08). DTU Compute.

Henze, M. & Comeau, Y. 2008 *Biological Wastewater Treatment: Principles Modelling and Design – Chapter 3, Wastewater Characterization*. IWA Publishing, London.

Henze Jr, M., Grady, C. P. L., Gujer, W., Marais, G. v. R. & Matsuo, T. 1987 *Activated Sludge Model no. 1*. Technical report, IAWPRC.

Henze, M., Gujer, W., Mino, T. & van Loosdrecht, M. C. M. 2000 *Activated Sludge Models: ASM1, ASM2, ASM2d and ASM3*. Scientific and Technical Report no. 9. IWA Publishing, London, UK.

Hindmarsch, A. C. 1983 *ODEPACK, A Systematized Collection of ODE Solvers*, 1st edn. Scientific Computing (IMACS Transaction on Scientific Computation), Amsterdam.

Hu, Z., Wentzel, M. C. & Ekama, G. A. 2007 A general kinetic model for biological nutrient removal activated sludge systems: model development. *Biotechnology and Bioengineering* **98** (6), 1242–1258.

Isaacs, S. H. & Thornberg, D. 1998 A comparison between model and rule based control of a periodic activated sludge process. *Water Science and Technology* **37** (12), 343–351.

Jazwinski, A. H. 1970 *Stochastic Processes and Filtering Theory*. Dover Publications, Inc., Mineola, New York.

Juhl, R., Møller, J. K., Jørgensen, J. B. & Madsen, H. 2016 Modeling and prediction using stochastic differential equations. In: *Prediction Methods for Blood Glucose Concentration. Lecture Notes in Bioengineering* (H. Kirchsteiger, J. Jørgensen, E. Renard & L. del Re, eds). Springer, Cham.

Kim, H., Kim, Y., Kim, M., Piao, W., Gee, J. & Kim, C. 2014 Performance evaluation of a full-scale advanced phase isolation ditch process by using real-time control strategies. *Korean J. Chem. Eng.* **31** (4), 611–618.

Kristensen, N. R. & Madsen, H. 2003 *Continuous Time Stochastic Modelling Mathematics Guide*.

Kristensen, N. R., Madsen, H. & Jørgensen, S. B. 2004 Parameter estimation in stochastic grey-box models. *Automatica* **40** (2), 225–237.

Lewis, W. & Whitman, W. 1924 Principles of gas absorption. *Industrial and Engineering Chemistry* **16** (12), 1215–1220.

Löwe, R., Vezzaro, L., Mikkelsen, P. S., Grum, M. & Madsen, H. 2016 Probabilistic runoff volume forecasting in risk-based optimization for RTC of urban drainage systems. *Environmental Modelling and Software* **80**, 143–158.

Meijer, S. C. F. 2004 *Theoretical and Practical Aspects of Modelling Activated Sludge Processes*. PhD Thesis, Delft University of Technology, The Netherlands.

Mulas, M., Tronci, S. & Baratti, R. 2007 Development of a 4-Measureable States Activated Sludge Process Model Deduced from the ASM1. In: *IFAC Proceedings vol. 1*, June 4–6, Cancún, Mexico.

Nielsen, M. K. & Önnerth, T. B. 1995 Improvement of a recirculating plant by introducing STAR control. *Water Science and Technology* **31** (2), 171–180.

Petersen, B., Vanrolleghem, P. A., Gernaey, K. & Henze, M. 2002 Evaluation of an ASM1 model calibration procedure on a municipal-industrial wastewater treatment plant. *Journal of Hydroinformatics* **4** (1), 15–38.

Pittoors, E., Guo, Y. & Van Hulle, S. W. H. 2014 Modeling dissolved oxygen concentration for optimizing aeration systems and reducing oxygen consumption in activated sludge processes: a review. *Chem. Eng. Comm.* **201**, 983–1002.

Rieger, L., Alex, J., Winkler, S., Boehler, M., Thomann, M. & Siegrist, H. 2003 Progress in sensor technology – progress in process control? Part 1: sensor property investigation and classification. *Water Science and Technology* **47** (2), 103–112.

Sharifi, S., Murthy, S., Takács, I. & Massoudieh, A. 2014 Probabilistic parameter estimation of activated sludge processes using Markov Chain Monte Carlo. *Water Research* **50**, 254–266.

Sin, G., De Pauw, D. J. W., Weijers, S. & Vanrolleghem, P. 2008 An efficient approach to automate the manual trial and error calibration of activated sludge models. *Biotechnology and Bioengineering* **100** (3), 516–528.

Smets, I. Y., Haegebaert, J. V., Carrette, R. & Van Impe, J. F. 2003 Linearization of the activated sludge model ASM1 for fast and reliable predictions. *Water Research* **37**, 1831–1851.

Sørensen, J., Thornberg, D. E. & Nielsen, M. K. 1994 Optimization of a nitrogen-removing biological wastewater treatment plant using on-line measurements. *Water Environment Research* **66** (3), 236–242.

Speelpenning, B. 1980 Compiling Fast Partial Derivatives of Functions Given by Algorithms. Technical Report UILU-ENG 80 1702, University of Illinois, Urbana, USA. doi: 10.2172/5254402.

Stentoft, P. A., Munk-Nielsen, T., Mikkelsen, P. S. & Madsen, H. 2017 A Stochastic Method to Manage Delay and Missing Values for In-Situ Sensors in an Alternating Activated Sludge. In: *Proceedings of ICA 2017 11-14 June 2017 in Québec City*, Québec, Canada.

Thordarson, F., Breinholt, A., Moller, J. K., Mikkelsen, P. S., Grum, M. & Madsen, H. 2012 Evaluation of probabilistic flow predictions in sewer systems using grey box models and a skill score criterion. *Stochastic Environmental Research and Risk Assessment* **26** (8), 1151–1166.

Tullekin, H. J. A. F. 1993 Grey-box modelling and identification using physical knowledge and Bayesian techniques. *Automatica* **29** (2), 285–308.

Vanrolleghem, P. A., Van Daele, M. & Dochain, D. 1995 Structural identifiability of biokinetic models of activated-sludge respiration. *Water Research* **29** (11), 2571–2578. https://doi.org/10.1016/0043-1354(95)00106-U.

Wentzel, M. C., Ekama, G. A. & Marais, G. v. R. 1992 Process and modelling of nitrification denitrification biological excess phosphorus removal systems – a review. *Water Science and Technology* **25** (6), 59–82.

Zhao, H., Freed, A. J., Dimassimo, R. W., Hong, S. N., Bundgaard, E. & Thomsen, H. A. 2004 *Demonstration of Phase Length Control of BioDenipho Process Using On-Line Ammonia and Nitrate Analyzers at Three Full-Scale Wastewater Treatment Plants*. WEFTEC, Alexandria, VA, USA.

First received 3 August 2018; accepted in revised form 17 December 2018. Available online 28 December 2018

Ammonia-based aeration control with optimal SRT control: improved performance and lower energy consumption

Oliver Schraa, Leiv Rieger, Jens Alex and Ivan Miletić

ABSTRACT

Ammonia-based aeration control (ABAC) is a cascade control concept for controlling total ammonia nitrogen (NH$_x$-N) in the activated sludge process. Its main goals are to tailor the aeration intensity to the NH$_x$-N loading and to maintain consistent nitrification, to meet effluent limits but minimize energy consumption. One limitation to ABAC is that the solids retention time (SRT) control strategy used at a water resource recovery facility (WRRF) may not be consistent with the goals of ABAC. ABAC-SRT control is a strategy for aligning the goals of ammonia-based aeration control and SRT control. A supervisory controller is used to ensure that the SRT is always optimal for ABAC. The methodology has the potential to reduce aeration energy consumption by over 30% as compared to traditional dissolved oxygen (DO) control. Practical implementation aspects are highlighted for implementation at full scale, such as proper selection of the set point for the supervisory controller, proper calculation of the rate of change in sludge inventory, using a mixed liquor suspended solids (MLSS) controller, and tuning of the controllers. In conclusion, ABAC-SRT is a promising approach for coordinated control of SRT, total ammonia nitrogen, and dissolved oxygen in the activated sludge process that balances both treatment performance and energy savings.

Key words | ammonia-based aeration control, energy, modeling, nutrient removal, simulation, SRT control

Oliver Schraa (corresponding author)
Leiv Rieger
Ivan Miletić
inCTRL Solutions Inc.,
107-7 Innovation Drive, Dundas, ON, L9H 7H9,
Canada
E-mail: *schraa@inctrl.ca*

Jens Alex
ifak e.V. Magdeburg,
Werner-Heisenberg-Str. 1, 39106 Magdeburg,
Germany

INTRODUCTION

Ammonia-based aeration control (ABAC; Rieger *et al.* 2014) is a cascade control concept for controlling total ammonia nitrogen (NH$_x$-N) in the activated sludge process. Its main goals are to tailor the aeration intensity to the NH$_x$-N loading and to maintain consistent nitrification, to meet effluent limits but minimize energy consumption and improve nutrient removal (Rieger *et al.* 2012).

One limitation to ABAC encountered by the authors is that the solids retention time (SRT) control strategy used at a water resource recovery facility (WRRF) may not be consistent with the goals of ABAC. For example, ABAC may not be able to handle peak loads if the SRT is too low and may reach minimum airflow constraints if the SRT is too high. In order to overcome this limitation, Schraa *et al.* (2016) introduced the concept of ABAC-SRT control, where an ABAC system is combined with a dynamic SRT controller. A higher-level supervisory controller is used to coordinate the two controllers. The supervisory controller

determines an optimal SRT set point that ensures the ammonia controller can meet its goals without encountering constraints. The objectives of the current study are to review the methods available for calculating SRT, to determine the most appropriate SRT calculation method in the context of feedback control, to introduce the ABAC-SRT control concept, to demonstrate the economic benefits of ABAC-SRT control, and to explore practical implementation aspects.

CONTROL CONCEPTS

SRT control

The SRT is an important design and control parameter for the activated sludge process. It represents the average length of time that bacteria stay in the process before

doi: 10.2166/wst.2019.032

being wasted or lost in the effluent. The selected SRT directly affects the bacterial composition in an activated sludge system. Maintaining an appropriate SRT is necessary to ensure that slow-growing bacteria, such ammonia-oxidizing organisms, are not washed out of the system.

In the wastewater treatment literature, SRT is usually considered from a steady-state perspective. At steady-state, the static SRT is defined as the mass of microorganisms in the system divided by the mass of microorganisms wasted per day:

$$SRT = \frac{X_{MLSS}V_a}{X_{RAS}Q_w + X_eQ_e} \tag{1}$$

where:

V_a = total volume of aeration tanks, m^3

X_{MLSS} = total or volatile suspended solids (VSS) concentration in the mixed liquor, mg/L

X_{RAS} = total or VSS concentration in the return activated sludge (RAS), mg/L

X_e = total or VSS concentration in the secondary effluent, mg/L

Q_w = waste activated sludge (WAS) flow rate, m^3/d

Q_e = effluent flow rate, m^3/d

As shown in Equation (1), the VSS or total suspended solids (TSS) concentrations are typically used as surrogate measurements for the microorganism concentrations. The static SRT is only equal to the actual SRT at steady-state because if the WAS flow rate is changed the formula suggests that the SRT has instantly changed, which is not true in reality because the system takes considerable time to respond. This limitation can be overcome by filtering the static SRT with a low-pass filter or calculating the dynamic SRT as demonstrated later in this paper. Equation (1) also ignores the active microorganisms in the clarifier sludge blanket, which could be significant but are difficult to measure accurately.

For a completely mixed activated sludge system, a mass balance on microorganisms can be re-arranged to calculate the specific substrate utilization rate, which is the mass of substrate utilized per day per mass of active microorganisms:

$$U = \frac{1 + b_hSRT}{Y(SRT)} \tag{2}$$

where:

U = specific substrate utilization rate, mass chemical oxygen demand (COD)/mass VSS/d

Y = activated sludge yield coefficient, mass VSS/mass COD

b_h = endogenous decay rate, d^{-1}

As shown in Equation (2), the specific substrate utilization rate is directly related to the yield, decay rate, and SRT. Both the activated sludge yield and decay rate are considered to be constant for practical purposes, leaving the SRT as the main parameter used to control the substrate utilization rate.

The SRT in a WRRF is typically controlled by adjusting the WAS flow rate. One especially important factor to consider in SRT control is the speed of response. A plant's SRT takes at least two to three times the steady-state SRT to stabilize after a change in waste flow. As a result, SRT control cannot remove the variations in SRT caused by the diurnal loading variations. These occur at too high a frequency to be attenuated by changing the waste flow rate. The main goals of an SRT controller are to maintain a consistent SRT, and to respond to seasonal variations and storm events.

Equation (1) can be simplified by assuming that the biomass lost in the effluent is negligible. Using this assumption, a common SRT control strategy is hydraulic wasting (Garrett 1958) where sludge is wasted directly from the aeration tanks and the desired waste flow rate becomes the volume of the aeration tanks divided by the desired SRT:

$$Q_w = \frac{V_a}{SRT} \tag{3}$$

When sludge is wasted from the recycle line, with recycle ratio r, the hydraulic wasting formula becomes (WEF 1997):

$$Q_w = \left(\frac{r}{r+1}\right)\frac{V_a}{SRT} \tag{4}$$

Stephenson *et al.* (1981) and Brewer *et al.* (1995) show practical applications of implementing hydraulic wasting strategies. Hydraulic wasting is simple to understand and execute, but one potential disadvantage is that for a constant SRT the waste flow does not vary and this may not be optimal during storm events and seasonal variations.

SRT can also be controlled with an automatic feedback control algorithm that uses on-line measurements of mixed liquor suspended solids (MLSS), recycled activated sludge total suspended solids (RAS TSS), and waste flow rate as suggested by Vaccari *et al.* (1988) and Ekster (2001). A potential algorithm could use the on-line measurements to calculate the SRT using Equation (1), filter the static SRT using a low-pass filter, and have the feedback controller adjust the waste flow rate to keep the filtered SRT at the set point. The concept is illustrated in Figure S1 in the Supplementary Material (available with the online version of this paper). The filtering of the SRT calculation (or the flowrate and TSS measurements) is used as the steady-state SRT responds too rapidly to dynamic process variations and set-point changes. Using a moving-average SRT is not ideal as its response is not smooth enough for proper feedback control.

Vaccari et al. (1985) developed a dynamic sludge age (DSA) function based on an age-balance equation in order to overcome the limitations of the static SRT. Analytical expressions for the DSA were developed by Vaccari et al. (1985) for four common cases. Takács & Patry (2002) and Takács (2008) developed the dynamic SRT (DSRT) which is the solution of the following ordinary differential equation:

$$\frac{dSRT}{dt} = 1 - \frac{SRT(F_p)}{M} \qquad (5)$$

where:

$\dfrac{dSRT}{dt}$ = age change of solids (change in age of solids [in days] per days of real time)

M = mass of solids in the system

F_p = mass flow of solids produced in the system (true sludge production)

The DSA of Vaccari et al. (1985) and the DSRT of Takács & Patry (2002) can be shown to be equivalent using simulation provided that the same definitions are used for model variables such as true sludge production. Takács (2008) estimated the true sludge production using a model presented by Dold (2007). Alternatively, true sludge production can be estimated using the following equation from Vaccari et al. (1988):

$$F_p = \frac{M - M_0}{\Delta t} + Q_w X_w + Q_e X_e \qquad (6)$$

where:

M = mass of solids in the system at the current time (g)

M_0 = mass of solids in the system at the previous time interval (g)

Δt = time interval between calculations of the sludge production (d)

Q_w = waste flow rate (m³/d)

X_w = TSS concentration of waste stream (g/m³)

Vaccari et al. (1988) proposed feedback proportional-integral-derivative (PID) control of the DSA, but their investigations did not consider how the SRT set point itself should be optimized. This is one of the main objectives of the current study and will be addressed in the section below on ammonia-based aeration control with optimal SRT control (ABAC-SRT).

Comparison of SRT calculation methods

A comparison of selected SRT calculation methods was performed using SIMBA# water, a dynamic simulator for WRRFs. The SRT calculations were compared using an example nitrifying WRRF presented by Ekama & Wentzel (2008). The model of the WRRF (Figure 1) includes diurnal influent flow and COD, ammonia and ammonium nitrogen (NH_x), and soluble phosphorus (SP) concentration patterns (Diurnal Influent block), primary clarifiers, three bioreactors in series, and a layered clarifier model (Secondary clarifiers block). A separate dissolved oxygen (DO) controller is used for each bioreactor and each DO controller sends an airflow set point to a lower level flowrate controller that manipulates a control valve (contained within the Ammonia and SRT Controllers block). A detailed aeration system sub-model is included as part of the simulation model (Schraa et al. 2017) to provide a realistic test environment (contained within the Ammonia and SRT Controllers block). The aeration system model includes three 4,800 Nm³/h (3,075 standard cubic feet per minute – scfm) turbo blowers, aeration piping including fittings (based on a typical aeration piping layout), and membrane disc diffusers. A total airflow controller adjusts the blower output and number of blowers in service to match the sum of the three airflow controller set points.

The diurnal patterns are created using the influent generation tool developed by the HSG group (Langergraber et al. 2008, 2009). The pattern is adjusted on the weekends so that the loadings are reduced by 10% and the patterns are delayed by 1 hour.

Proportional recycle is used and wasting is done from the third bioreactor to provide an SRT of 11 days at design conditions. The SRT of 11 days was calculated as the minimum SRT for nitrification at 10 °C, a desired steady-state effluent NH_x concentration of 1 mgN/L, and a safety factor of 1.5 days using the design equation presented by van Haandel & van der Lubbe (2012).

The model of the example WRRF developed in SIMBA# uses the inCTRL-ASM biokinetic model along with the Otterpohl & Freund (1992) model for the primary clarifier and the Takács et al. (1991) clarification model with 10 layers for the secondary clarifiers. The Takács et al. (1991)

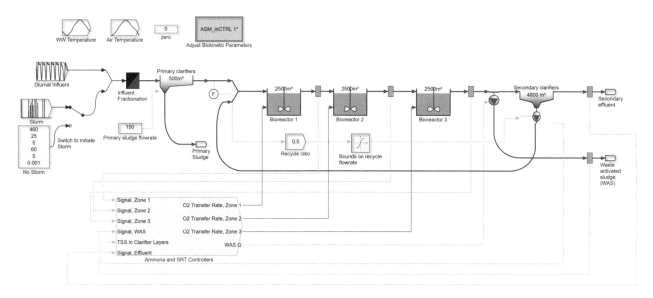

Figure 1 | Example WRRF as represented in SIMBA#.

model is used as it can be calibrated to predict an increase in the solids lost in the effluent as the flowrate increases. This is important in the current study as the SRT calculation is affected by the loss of solids in the clarifier effluent and by movement of solids from the aeration basins to the clarifier due to high flows. Although deficiencies in the Takács *et al.* (1991) model have been identified (Plósz *et al.* 2011; Bürger *et al.* 2013), the model is thought to be adequate for the current study as we are interested in the relative changes in the effluent TSS in response to changes in plant flows and loads and whether our proposed control concept can adapt to these changes. When implemented in practice, the SRT calculations will use actual TSS measurements.

The SRT calculation methods explored are as follows:

- Dynamic SRT calculation (Equations (5) and (6))
- Static SRT calculation (Equation (1))
- Filtered Static SRT and a filter time constant of 7 days (Equation (1))
- Hydraulic SRT (Equation (3) re-arranged to solve for SRT).

Dynamic simulations were conducted to compare the response of the SRT calculations to a step change in WAS flow rate and to a storm event at a wastewater temperature of 10 °C. The simulations were initialized by first running a 100-day simulation with the diurnal loading pattern at 10 °C to achieve a stable operating point, and then 60-day diurnal simulations were conducted to study the impact of a WAS flow change and a storm event.

A plot of the response of the different SRT calculation methods to a diurnal influent loading pattern with a step change in WAS flow rate at 5 days is shown in Figure 2. The hydraulic SRT starts at 11 days, while the other SRT values average 10.5 days as they consider the solids lost in the effluent. As shown in Figure 2, the hydraulic SRT and the static SRT are very sensitive to changes in the WAS flow and instantly move to the new SRT. An instantaneous increase in SRT is unrealistic as the SRT cannot increase by more than 1 day per day with no wasting. The dynamic SRT responds much more smoothly and slowly to the change in WAS flowrate and is better suited to automatic SRT control. The filtered static SRT is a reasonable approximation to the dynamic SRT for changes in WAS flow, but the filter time constant used becomes a tuning parameter and would need to be varied depending on the desired SRT set point.

In the Supplementary Material (available online), Figure S2 shows a plot of the response of the different SRT calculation methods to a diurnal influent loading pattern with a storm event that occurs after 10 days with a constant wasting rate. It is found that the hydraulic and traditional SRT calculation methods differ considerably from the dynamic SRT during a storm. The hydraulic SRT stays constant throughout the entire simulation because the wasting rate does not change. The static SRT calculation shows a large drop in the SRT during the storm, due to the loss of solids in the effluent, but then quickly returns back to its original value, which is not possible because the SRT cannot increase by more than 1 day per day. Clearly, the hydraulic and static SRT calculations are unrealistic under storm conditions.

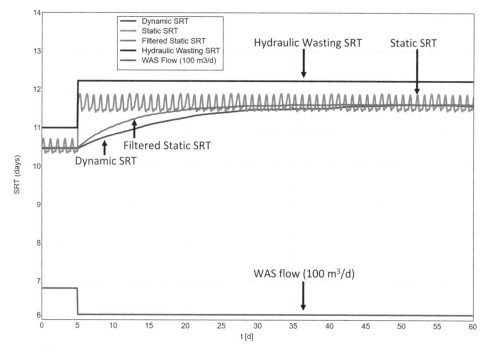

Figure 2 | Comparison of different SRT calculations after a step change in WAS flow rate.

The dynamic SRT and the filtered static SRT both drop in response to a storm but do so more slowly and then take much longer to return to the original SRT. Clearly, the dynamic SRT and the filtered static SRT are more realistic and appropriate for feedback control than the traditional or hydraulic wasting SRT calculations as they have a smoother dynamic response with the correct rate of change.

Ammonia-based aeration control (ABAC)

As discussed earlier, ABAC is a cascade control concept for controlling total ammonia nitrogen (NH_x-N) in the activated sludge process. A diagram illustrating the concept of ABAC is shown in Figure 3.

Ammonia-based aeration control with optimal SRT control (ABAC-SRT)

The ABAC-SRT control concept developed by Schraa *et al.* (2016) is shown in Figure 4. A feedback controller measures NH_x-N and manipulates the set point of a DO controller that manipulates an air flowrate set point for an airflow controller. A supervisory controller manipulates the set point of the SRT controller to ensure that the ammonia controller can achieve its goals without encountering constraints on minimum airflows or the nitrifier population. The supervisory controller is a feedback controller that controls the average DO concentration calculated by the ammonia controller. Averaging of the DO set point calculated by the ammonia controller is performed using a low-pass filter.

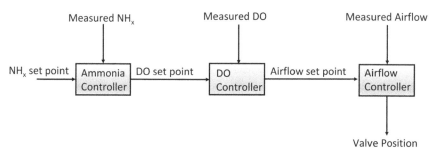

Figure 3 | Control concept for ammonia-based aeration control (ABAC).

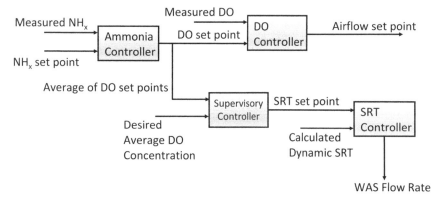

Figure 4 | Control concept for ammonia-based aeration control combined with optimal SRT control (ABAC-SRT).

The SRT controller controls the estimated dynamic SRT (Takács & Patry 2002) by manipulating the WAS flow rate.

Another important consideration is the selection of the SRT controller output bounds. At low WAS flows, the resulting MLSS could be too high and cause clarifier failure. To ensure that the MLSS stays in an acceptable range, the SRT controller cascades to an MLSS controller that has bounds on the MLSS set point, as shown in Figure S3 in the Supplementary Material (available online). Bounds are also placed on the SRT set point calculated by the supervisory controller so that the SRT does not become too low or too high (see Table 1).

RESULTS AND DISCUSSION

Comparison of control strategies

A case study of the ABAC-SRT concept was conducted by Schraa *et al.* (2016) but did not reveal the full benefits of the methodology as the practical aspects of selection of the average DO set point for the supervisory controller and proper selection of controller bounds had not been fully developed. The objective here is to demonstrate the potential benefits while providing guidance on practical implementation issues.

Table 1 | Summary of simulated scenarios for comparing performance of DO control, ABAC, and ABAC-SRT

Scenario	Airflow control	DO control	NH_x-N control	MLSS control	SRT control	Supervisory control
Case 1	- Airflow set point provided by DO controller for each zone	- DO set point of 2 mg/L in all zones	- Not used	- MLSS set point = 2,000 mg/L - WAS flow rate bounded between 100 and 3,500 m³/d	- Not used	- Not used
Case 2	- Airflow set point provided by DO controller for each zone	- DO set point provided by NH_x-N controller - Airflow set point bounded between minimum airflow for mixing and maximum airflow per diffuser	- NH_x-N set point = 1 mg/L - DO set point bounded between 0.5 and 2 mg/L	- MLSS set point = 2,000 mg/L - WAS flow rate bounded between 100 and 3,500 m³/d	- Not used	- Not used
Case 3	- Airflow set point provided by DO controller for each zone	- DO set point provided by NH_x-N controller - Airflow set point bounded between minimum airflow for mixing and maximum airflow per diffuser	- NH_x-N set point = 1 mg/L - DO set point bounded between 0.5 and 2 mg/L	- MLSS set point provided by SRT controller - WAS flow rate bounded between 100 and 3,500 m³/d	- SRT set point provided by supervisory controller - MLSS set point bounded between 1,000 and 3,500 mg/L	- Average DO set point of 1 mg/L - SRT set point bounded between 3 and 20 days

The ABAC-SRT control concept was implemented in SIMBA# and applied to the example WRRF introduced earlier. A year-long simulation is conducted in SIMBA# with 12 storm events and seasonal temperature variation (air and wastewater). The wastewater temperature varies between 12.4 and 24 °C, and the air temperature varies between −7.1 °C and 23.6 °C based on data taken from a wastewater treatment plant in Ontario, Canada.

To demonstrate the benefits of ABAC-SRT control, three cases are compared: Case 1 – DO control with MLSS control, Case 2 – ABAC with MLSS control, and Case 3 – ABAC-SRT. For the DO control and ABAC options, the MLSS is controlled using a proportional-integral (PI) controller with a set point of 2,000 mg/L. MLSS control is used in Cases 1 and 2 as it is thought to represent a common control strategy at many WRRFs due to concerns about clarifier failure at high MLSS concentrations. The supervisory controller has minimum and maximum output bounds of 3 days and 20 days respectively, to impose bounds on the SRT set point. The SRT controller has MLSS bounds of between 1,000 and 3,500 mg/L. The simulated scenarios are summarised in Table 1.

In Case 1, the NH_x-N is uncontrolled and over-aeration is possible. In Cases 2 and 3, the NH_x-N set point in the third bioreactor is set to 1 mgN/L in order to achieve potential energy savings. The airflow controllers are bounded between 0.22 Nm^3/h per m^2 of floor area (0.12 scfm/ft^2

based on USEPA (1989)) and 14 Nm^3/h per diffuser (7.6 scfm/diffuser) to ensure adequate airflow for mixing and to remain within the upper airflow limit for the diffusers. The DO set point calculated by the ammonia controller is bounded between 0.5 mg/L (to prevent extended operation at low DO concentrations in cases without SRT control) and 2 mg/L (to prevent excessive aeration). The ammonia controller is a PID controller and all the remaining controllers are PI controllers.

The year-long influent flowrate and air and wastewater temperature patterns are shown in Figure 5. Figure 6 shows the resulting simulated dynamic SRT (controlled variable) and the WAS flow rate (manipulated variable) for Case 1 – DO control with MLSS control, Case 2 – ABAC with MLSS control, and Case 3 – ABAC-SRT. As shown, in Case 3 the WAS flow rate is adjusted so that the SRT varies throughout the year. The SRT variation follows the wastewater temperature variation. The lower SRT and MLSS bounds of 3 days and 1,000 mg/L respectively become active during the warmer months.

Figure 7 shows the load-based average NH_x-N in the third bioreactor (controlled variable) and the load-based average DO in the first bioreactor (manipulated variable). In all three cases, the ammonia controller cannot keep the daily average Zone 3 NH_x-N at the set point of 1 mgN/L at all times. In Case 1, this is because a DO of 2 mg/L is maintained in all three zones, which is sufficient for full

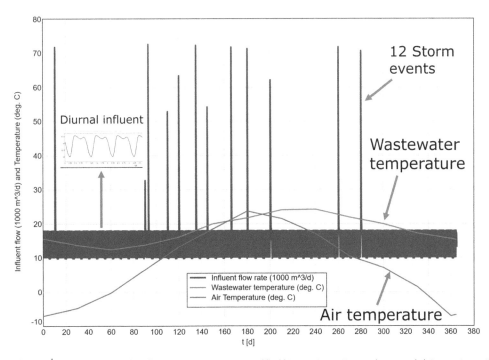

Figure 5 | Diurnal flow pattern including 12 storm events, a seasonal liquid temperature pattern, and a seasonal air temperature pattern for the 365-day simulation.

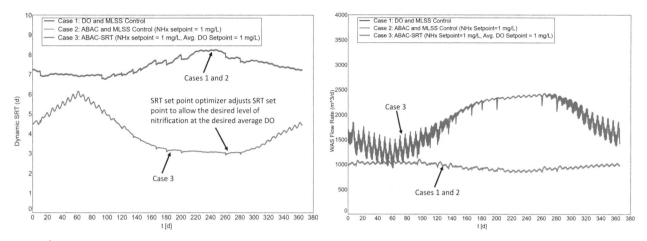

Figure 6 | Simulated dynamic SRT (left plot) and WAS flow rate (right plot) for DO and MLSS control, ABAC and MLSS control, and ABAC-SRT control.

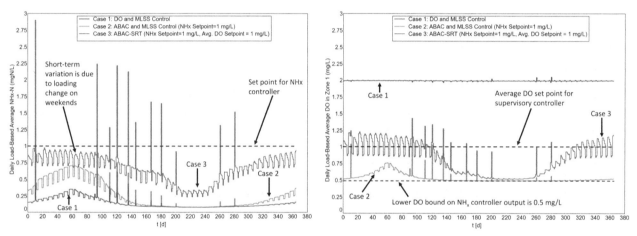

Figure 7 | Simulated load-based average NH_x-N in third bioreactor (left plot) and Zone 1 DO (right plot) for DO and MLSS control, ABAC and MLSS control, and ABAC-SRT control.

nitrification at the operating SRT (7 days and above). In Case 2, the ammonia controller reaches its lower DO bound of 0.5 mg/L, but the DO in the last zone is consistently above 2 mg/L because of the minimum airflow bound for mixing. At the SRTs encountered, this is enough to keep the daily average Zone 3 NH_x-N significantly below the 1 mgN/L set point. The ABAC-SRT controller performs best in terms of achieving the NH_x-N set point because of the supervisory controller.

In Case 3 with ABAC-SRT, the goal of the supervisory controller was to keep the average DO at 1 mg/L. This is achieved during the colder months but the average DO drops lower than 1 mg/L in the warmer months because of the minimum SRT bound of 3 days. The minimum SRT bound also causes the ammonia controller to be limited by its lower DO set point bound of 0.5 mg/L, which in turn allows the daily average Zone 3 NH_x-N to drop much lower than its set point during warmer months.

The ABAC-SRT control concept ensures that the SRT is long enough to attenuate peak loads while at the same time ensuring that aeration system constraints do not always limit control authority. Despite this, ABAC-SRT can still be limited to some extent by the bounds on SRT, MLSS, and DO set points, which provide a measure of safety. These set point bounds can be fine-tuned to allow unconstrained operation of the control system if desired.

Aeration energy consumption was calculated by the model for the three cases and was 3,260 kWh/d for DO and MLSS control, 2,550 kWh/d for ABAC with MLSS control, and 2,220 kWh/d for ABAC-SRT control. See Figure S4 in the Supplementary Material (available with the online version of this paper) for the cumulative energy consumption for the entire simulation for the three cases. The ABAC with MLSS control strategy results in a 22% reduction in energy consumption as compared to DO and

MLSS control. This energy reduction is consistent with that reported by Amand *et al.* (2013) and Rieger *et al.* (2012) for full-scale WRRFs that implemented ABAC.

The ABAC-SRT control strategy results in a 32% reduction in energy consumption as compared to DO and MLSS control, an additional 10% in savings due to SRT set-point optimisation. Another potential benefit of the ABAC-SRT strategy is a lower MLSS concentration, which helps minimize solids loss during storms for plants operated at high MLSS concentrations. As shown, the level of savings with ABAC and ABAC-SRT depends on the NH_x-N set point. They are also impacted by the lower bound on the DO set point and, in the case of ABAC-SRT, by the lower bound on the SRT and MLSS set points.

Controller implementation issues

An important consideration when implementing ABAC-SRT is the selection of the average DO set point of the supervisory controller. In the case study discussed earlier, the average DO set point of the supervisory controller was 1 mg/L. The average DO set point determines whether the SRT is in the correct range so that the ammonia controller can function properly to prevent NH_x-N break-through and to ensure that energy savings are maximized. See Figure S5 in the Supplementary Material (available online) for the impact on the controlled NH_x-N when the average DO set point is changed to 0.6 mg/L.

Another implementation issue is the selection of the time interval used in Equation (6). Longer time intervals serve to filter the impact of diurnal variations on the calculation of the sludge production and the impact of solids being pushed into the clarifiers during storms. If Δt is too small, the dynamic SRT will vary considerably in response to diurnal variations and will be incorrectly impacted by solids being pushed into the clarifiers during a storm (as the solids are temporarily lost as far as the calculation is concerned unless the solids mass in the clarifier is tracked). A time interval of 1 day in Equation (6) was found to provide a reasonable compromise between accuracy and eliminating the negative impact of high frequency flow variations.

Proper controller tuning is also very important for the ABAC-SRT control strategy. The system contains a number of cascaded controllers that could exhibit poor performance if not tuned properly. The lowest loops in the cascade (i.e. airflow controllers and the MLSS controller) should be tuned for a faster response than the loops higher in the cascade. As a starting point for controller tuning,

correlations such as Ciancone & Marlin (1992) can be used. Use of tuning correlations requires knowledge of the process gain, time constant, and transportation lag for each control loop, which can be determined using a simulation model or plant step-response tests.

In the SIMBA# model, the controllers were fine-tuned by introducing step changes in the controller set points and adjusting the tuning constants to achieve the desired dynamic response. The MLSS and SRT controllers were tuned for a slow dynamic response to ensure that the WAS flow variations were not overly aggressive. The controller integral times are on the order of days for the SRT-related controllers and on the order of minutes for the ammonia, DO, and air flowrate controllers. The SRT-related controller gains depend on the SRT range that the system is operating within, suggesting that gain scheduling could be beneficial.

CONCLUSIONS

ABAC-SRT control is a strategy for aligning the goals of ammonia-based aeration control and SRT control. A supervisory controller is used to ensure that the SRT is always optimal for ammonia-based aeration control. The methodology has the potential to reduce aeration energy consumption by over 30% as compared to traditional DO control. Practical implementation aspects were highlighted for implementation at full scale, such as proper selection of the average DO set point for the supervisory controller, selection of the time interval used in calculating the rate of change in sludge inventory, using an MLSS controller to ensure that the SRT controller does not lead to excessively low or high MLSS concentrations, and proper tuning of the controllers. In conclusion, the ABAC-SRT concept is a promising approach for coordinated control of SRT, total ammonia nitrogen, and DO in the activated sludge process, which balances both treatment performance and energy savings.

DECLARATION

Certain portions of the material described in this paper are patent pending and the submission and/or publication of this paper does not grant any license or other right in respect of any intellectual property owned by the authors, inCTRL Solutions Inc., Institut für Automation und Kommunikation e.V. or any related entities.

REFERENCES

Amand, L., Olsson, G. & Carlsson, B. 2013 Aeration control – a review. *Wat. Sci. & Tech.* **67** (11), 2374–2398.

Brewer, H. M., Stephenson, J. P. & Green, D. 1995 Plant Optimization Using Online Phosphorus Analyzers and Automated SRT Control to Achieve Harbour Delisting. In: *Proceedings of WEFTEC 1995*, Miami, FL.

Bürger, R., Diehl, S., Farås, S., Nopens, I. & Torfs, E. 2013 A consistent modelling methodology for secondary settling tanks: a reliable numerical method. *Wat. Sci. & Tech.* **68** (1), 192–208.

Ciancone, R. & Marlin, T. 1992 Tune controllers to meet your performance goals. *Control* **5**, 50–57.

Dold, P. L. 2007 Quantifying Sludge Production in Municipal Treatment Plants. In: *Proceedings of WEFTEC 2007*, October 13–17, San Diego, CA.

Ekama, G. A. & Wentzel, M. C. 2008 *Biological Wastewater Treatment: Principles, Modelling and Design*. IWA Publishing, London, UK, Ch. 4 and 5.

Ekster, A. 2001 Automatic Waste Control. In: *Proceedings of the 1st IWA International Conference on Instrumentation, Control and Automation (ICA)*, June 3–7, Malmö, Sweden.

Garrett, M. T. 1958 Hydraulic control of activated sludge growth rate. *Sew. Ind. Wastes* **39** (3), 253–261.

Langergraber, G., Alex, J., Weissenbacher, N., Woerner, D., Ahnert, M., Frehmann, T., Halft, N., Hobus, I., Plattes, M., Spering, V. & Winkler, S. 2008 Generation of diurnal variation for dynamic simulation. *Wat. Sci. & Tech.* **59** (9), 1483–1486.

Langergraber, G., Spering, V., Alex, J., Ahnert, M., Cernochoca, L., Dürrenmatt, D. J., Frehmann, T., Hobus, I., Weissenbacher, N., Winkler, S. & Yücesoy, E. 2009 Using numerical simulation to optimize control strategies during activated sludge plant design. In: *Proceedings of the 10th IWA Conference on Instrumentation, Control and Automation (ICA)*, June 14–17, Cairns, Australia.

Otterpohl, R. & Freund, M. 1992 Dynamic models for clarifiers of activated sludge plants with dry and wet weather flows. *Wat. Sci. & Tech.* **26** (5–6), 1391–1400.

Plósz, B. G., Nopens, I., De Clercq, J., Benedetti, L. & Vanrolleghem, P. A. 2011 Shall we upgrade one-dimensional secondary settler models used in WWTP simulators? – an assessment of model structure uncertainty and its propagation. *Wat. Sci. & Tech.* **63** (8), 1726–1738.

Rieger, L., Takács, I. & Siegrist, H. 2012 Improving nutrient removal while reducing energy use at three Swiss WWTPs using advanced control. *Water Environ. Res.* **84** (2), 171–189.

Rieger, L., Jones, R. M., Dold, P. L. & Bott, C. B. 2014 Ammonia-based feedforward and feedback aeration control in activated sludge processes. *Water Environ. Res.* **86** (1), 63–73.

Schraa, O., Rieger, L. & Alex, J. 2016 Coupling SRT control with aeration control strategies. In: *Proceedings of WEFTEC.16*, New Orleans, LA, USA.

Schraa, O., Rieger, L. & Alex, J. 2017 Development of a model for activated sludge aeration systems: linking air supply, distribution, and demand. *Wat. Sci. & Tech.* **75** (3), 552–560.

Stephenson, J. P., Monaghan, B. A. & Laughton, P. J. 1981 Automatic control of solids retention time and dissolved oxygen in the activated sludge process. *Wat. Sci. & Tech.* **13**, 751–758.

Takács, I. 2008 *Experiments in Activated Sludge Modelling*. PhD Thesis, Ghent University, Belgium, pp. 267.

Takács, I. & Patry, G. G. 2002 The dynamic solids residence time. In: *Proceedings of IWA World Water Congress 2002*, Melbourne, Australia.

Takács, I., Patry, G. G. & Nolasco, D. 1991 A dynamic model of the thickening/clarification process. *Wat. Res.* **25** (10), 1263–1271.

U.S. Environmental Protection Agency (USEPA), Office of Research and Development 1989 *Design Manual: Fine Pore Aeration Systems*. EPA/625/1-89/023, U.S. E.P.A., Cincinnati, OH.

Vaccari, D. A., Fagedes, T. & Longtin, J. 1985 Calculation of mean cell residence time for unsteady-state activated sludge systems. *Biotech. Bioeng.* **27**, 695–703.

Vaccari, D. A., Cooper, A. & Christodoulatos, C. 1988 Feedback control of activated sludge waste rate. *Journal WPCF* **60**, 1979–1985.

van Haandel, A. C. & van der Lubbe, J. G. M. 2012 *Handbook of Biological Wastewater Treatment: Design and Optimisation of Activated Sludge Systems*. IWA Publishing, London.

WEF 1997 *Automated Process Control Strategies*. Water Environment Federation, Alexandria, VA, USA.

First received 19 October 2018; accepted in revised form 9 January 2019. Available online 23 January 2019

Tanks in series versus compartmental model configuration: considering hydrodynamics helps in parameter estimation for an N_2O model

Giacomo Bellandi, Chaïm De Mulder, Stijn Van Hoey, Usnam Rehman, Youri Amerlinck, Lisha Guo, Peter A. Vanrolleghem, Stefan Weijers, Riccardo Gori and Ingmar Nopens

ABSTRACT

The choice of the spatial submodel of a water resource recovery facility (WRRF) model should be one of the primary concerns in WRRF modelling. However, currently used mechanistic models are limited by an over-simplified representation of local conditions. This is illustrated by the general difficulties in calibrating the latest N_2O models and the large variability in parameter values reported in the literature. The use of compartmental model (CM) developed on the basis of accurate hydrodynamic studies using computational fluid dynamics (CFD) can take into account local conditions and recirculation patterns in the activated sludge tanks that are important with respect to the modelling objective. The conventional tanks in series (TIS) configuration does not allow this. The aim of the present work is to compare the capabilities of two model layouts (CM and TIS) in defining a realistic domain of parameter values representing the same full-scale plant. A model performance evaluation method is proposed to identify the good operational domain of each parameter in the two layouts. Already when evaluating for steady state, the CM was found to provide better defined parameter ranges than TIS. Dynamic simulations further confirmed the CM's capability to work in a more realistic parameter domain, avoiding unnecessary calibration to compensate for flaws in the spatial submodel.

Key words | ASMG2d, greenhouse gas emissions, mixing model, model layout, parameter domain

Giacomo Bellandi (corresponding author)
Department of Civil and Environmental Engineering, Environmental Section,
Polytechnic of Milan,
Piazza L. da Vinci, 32, 20133 Milan,
Italy
E-mail: Giacomo.Bellandi@polimi.it

Chaïm De Mulder
Youri Amerlinck
Ingmar Nopens
BIOMATH, Department of Mathematical Modelling, Statistics and Bioinformatics,
Ghent University,
Coupure Links 653, B-9000 Gent,
Belgium

Stijn Van Hoey
INBO, Herman Teirlinckgebouw,
Havenlaan 88 bus 73, 1000 Brussel (Anderlecht),
Belgium

Usnam Rehman
AM-TEAM,
Hulstbaan 63, 9100 Sint Niklaas,
Belgium

Lisha Guo
Ryerson University,
350 Victoria St, Toronto M5B 2K3, ON,
Canada
and
Trojan Technologies,
3020 Gore Road, London N5 V 4T7, ON,
Canada

Lisha Guo
Peter A. Vanrolleghem
modelEAU, Université Laval,
1065, avenue de la Médecine,
Québec G1 V 0A6, QC,
Canada

Stefan Weijers
Waterschap De Dommel,
Bosscheweg 56, 5283 WB Boxtel,
The Netherlands

Riccardo Gori
Department of Civil and Environmental Engineering,
University of Florence,
via di S. Marta 3, 50139 Florence,
Italy

doi: 10.2166/wst.2019.024

INTRODUCTION

Nitrous oxide (N_2O) has a GWP 265–298 times that of CO_2 for a 100-year timescale (IPCC 2013). Its emissions are also of great concern for water resource recovery facilities (WRRFs) (*inter alia:* Ravishankara *et al.* 2009) and modelling tools have been largely used in order to understand its production and define possible reduction strategies. The heterotrophic denitrification pathway model of Hiatt & Grady (2008) is currently the only generally accepted model. However, the pathways responsible for N_2O production are different and contribute to different extents to the emission depending on wastewater characteristics, plant dynamics and environmental conditions (*inter alia:* Daelman *et al.* 2015). Especially in full-scale applications, modelling is a fundamental tool for understanding N_2O production and emission dynamics. Mechanistic models have been applied to define general operational recommendations aimed at N_2O reduction (Ni & Yuan 2015) but still case-specific recommendations are necessary and more in depth process understanding is needed for an effective minimization of emissions.

A number of kinetic N_2O models describing very detailed biological processes have recently been developed (*inter alia:* Ni & Yuan 2015). In particular, models describing both ammonia oxidizing bacteria (AOB) pathways (i.e. AOB denitrification and incomplete hydroxylamine oxidation) have shown important advances in unfolding the contribution to N_2O production of the different consortia in laboratory controlled conditions (*inter alia:* Spérandio *et al.* 2016). These highly descriptive mechanistic models have been calibrated and validated in laboratory controlled conditions. However, despite the suggestion of Ni *et al.* (2013b) for using the dual pathway AOB models, Ni *et al.* (2013a) discouraged this implementation due to the risk of over-parametrization of the model and the possible creation of strong parameter correlations. In addition to this, the application of both dual pathway and single pathway models in full-scale is still troublesome due to recognized difficulties in identifying proper parameter sets (Ni *et al.* 2013b; Spérandio *et al.* 2016). Spérandio *et al.* (2016) observed high variability of different parameters, among the different case studies and the different models applied, with related high influence on N_2O and NO emission results (e.g. the AOB reduction factor set to a high value making the half saturation index for free nitrous acid (K_{FNA}) poorly identifiable). These large variations of parameters from one system to another are likely the result of concurring reasons, e.g.

micro-organisms' adaptation to local environmental and process conditions, defaults in the structure of the models, undescribed local heterogeneities in reactors (Spérandio *et al.* 2016).

The large variations of parameter values among different full-scale case studies considerably limit the predictive power of the models, as parameters cannot be extrapolated to other plants, and probably not even for different periods in the same plant. This also reduces the effective search for greenhouse gas emission mitigation strategies. Given the detailed mechanistic structure of available N_2O models, the considerable differences in parameters values among different (full-scale) applications can be explained by an unrealistic representation of local conditions in activated sludge (AS) tanks, to which these models are much more sensitive than traditional activated sludge model (ASM) processes.

The design of proper water resource recovery facility (WRRF) layouts (with respect to their spatial submodel) is an important step in plant-wide modelling and for understanding complex process dynamics (e.g. N_2O production) (Rehman *et al.* 2014). In current tanks in series (TIS) configurations, local recirculations and concentration gradients are assumed to be negligible, and the use of plug-flow/completely-mixed tank configurations is preferred to reduce overall model complexity and computational demand. To date, it is necessary to analyze the possibility and effect of the inclusion of more detailed descriptions of local concentrations in AS tanks by means of more detailed spatial submodels. The development of layouts designed to more accurately describe the hydrodynamic behavior of the internal volume can result in improved predictive power of available mechanistic models, which is key in optimization and control. Currently, the use of compartmental models (CMs) developed upon detailed computational fluid dynamics (CFD) studies is gaining interest from the modelling community (Le Moullec *et al.* 2010; Rehman *et al.* 2017, 2015, 2014).

In this work, a comparison of the performance of a CM and a TIS spatial submodel of the same full-scale WRRF on identifying a domain of good parameters values for the most influential parameters using the ASMG2d model (Guo 2014; Guo & Vanrolleghem 2014) is provided. Based on literature, each model parameter was sampled in a specific range for generating a number of simulation scenarios, each with a different parameter set. Each simulation scenario was ranked for its performance in predicting measured variables based on different criteria suggested by Van Hoey (2016). The latter returns the good performing scenarios in the form of a distribution of parameter values for both the CM and TIS.

MATERIALS AND METHODS

Model layouts

Two model layouts of the WRRF of Eindhoven were used, differing in terms of spatial submodel (Figure 1). The TIS layout of the Eindhoven WRRF (Figure 1, top) is a well consolidated model obtained after years of research of the facility (De Keyser *et al.* 2014; Amerlinck 2015). On the other hand, the CM version (Figure 1, bottom) is a recent development of the WRRF model layout resulting from a thorough hydrodynamic study based on CFD simulations in a three-phase (i.e. gas, solid, liquid) model integrated with an ASM to describe the biological activity (Rehman 2016). In particular, the volumes in which the biological tank was initially divided for the case of the TIS, were further partitioned by means of the cumulative species distribution concept that led to the development of the compartmental network currently in use.

For comparing the two model layouts, a common mechanistic reaction model was chosen with which comparison of the results was performed. Given the important efforts deployed to calibrate the ASMG1 and ASMG2d

models on the same plant, the biokinetic model chosen for this work was the ASMG2d (Guo 2014).

For the comparison of the two model layouts, three fundamental steps were followed: (I) parameter selection and sensitivity ranking; (II) steady state simulation of n-sampled parameter sets to confirm or redefine current parameter ranges; (III) dynamic simulations of n-sampled parameter sets to evaluate whether CM can better define the parameter domain than TIS. Throughout steady state and dynamic simulations, 12 different model fit metrics (see below) were assessed to evaluate the quality of the model output (see Supplementary Material (SM), available with the online version of this paper).

Parameter selection and sensitivity ranking (Step I)

A literature selection of the most influencing parameters for N_2O production contained in ASMG2d was performed. Screening the literature, a first set of 25 most uncertain parameters was selected (Gernaey & Jørgensen 2004; Hiatt 2006; Van Hulle *et al.* 2012; Mampaey *et al.* 2013; Ni *et al.* 2013b; Guo 2014; Spérandio *et al.* 2016) and is reported in Table 1. Some of the parameters show up to 140% deviation in different calibration exercises (Spérandio *et al.* 2016).

In order to ensure a sampling of the entire domain without excluding the maximum and minimum limits of each parameter, the domains reported in Table 1 were enlarged by 10% of the difference between the relative maximum and minimum values.

A global sensitivity analysis (GSA) was performed on this set of parameters using the Latin Hypercube – one

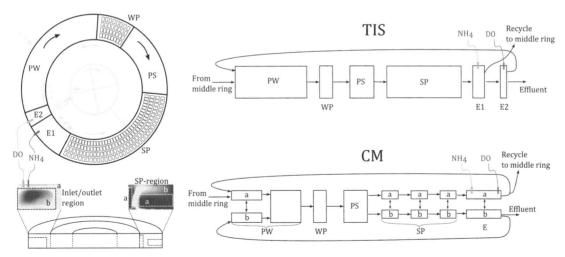

Figure 1 | Schematic representation of the partitioning of the AS tank volume according to the TIS (top) and CM (bottom) layouts. The planar representation of the AS tank (top left) is divided for the TIS (top right) in, so-called, pre-winter (PW), winter package (WP), pre-summer (PS), summer package (SP), effluent (E1 and E2) zones. The CM follows the same concept of TIS in the general division of the volumes, but includes *a* and *b* recirculation zones according to Rehman (2016).

sample at time (LH-OAT) approach (van Griensven *et al.* 2006) with different perturbation factors. As the choice of the perturbation factor can have an important effect on the numerical stability and thus on the sensitivity results, different magnitudes were investigated, i.e. from 10^{-2} to 10^{-6} (De Pauw & Vanrolleghem 2006). Also, the impact of the number of samples was observed in order to check whether the increase of one or two orders of magnitude impacted the final ranking. These tests resulted in consistent ranking of the outputs, with the only exception being the tests with the perturbation factors smaller than 10^{-5}, which resulted in numerical instabilities.

Simulations process

By means of a LH-OAT sampling approach on the most influential parameters resulting from Step I, the scenarios for the analysis in Step II and III were created. For Step I 10,000 simulations were run given the suggested minimum sample size in the parameter space (Van Hoey 2016), while in Steps II and III, each parameter was uniformly sampled in 2,000 points of its domain.

Step II

Steady state simulations were used to compare the model output concentrations with known normal operation conditions in the biological tank. Steady state simulations of 100 days were run, using as input data averaged measurements collected over two months of well-known good plant operation in dry conditions in the summer of 2012 with 1 min frequency. The output of the simulations was compared against average typical concentrations of ammonia (NH_4), dissolved oxygen (DO) and total suspended solids (TSS) at the end of the summer package aeration compartment (1.01 mg NH_4-N/L, 1.02 mgO$_2$/L and 3,200 gTSS/m^3, respectively). This allowed making a first ranking of the scenarios based on the proximity of the model outputs and the known measured values of NH_4, DO and TSS. As a result, this allowed evaluation of the domain of each parameter considered and eventually provide adjustments by repeating the steady state simulations. This iterative approach allowed definition of a domain for each parameter with 'good' parameter values, so that no possibly good parameters values were left out and, at the same time, exclude zones of undoubtedly bad parameter values in order to proceed with Step III. Results of Step II are mostly reported in SM.

Step III

Once the last parameter domains after the steady state were defined, the LH-OAT sampling on 2,000 points was repeated for creating the scenarios for the dynamic simulations. Parameters were uniformly sampled from the reduced parameter domain after the steady state analysis. One day of validated influent data with minute frequency was used as input to the model. The outputs were compared with a one minutes frequency dataset of measured of DO, NH_4, nitrate (NO_3) and liquid N_2O (Unisense Environment, Denmark).

Scenario ranking using 12 different metrics

Different metrics can be used to score a model fit according to a variety of methods assessing the similarity between a modelled and a measured dataset. The dissimilarity between the different metrics depends not only on their mathematical structure but also on the system behavior and the modelling objective (e.g. average and peak behavior). Hence, there is the need for an assortment of criteria to evaluate the performance of a model from different perspectives. For instance, the root mean square error (RMSE) is a commonly chosen metric to evaluate a model fit; however, it emphasizes the fit of peaks and extreme values. Therefore, its combination with RVE, from the total relative error category, is advisable when variables with a wide range of values are compared (Hauduc *et al.* 2015).

In this view, for both the steady state and the dynamic simulation step, the outputs were evaluated by means of 12 metrics, i.e. MAE, RMSE, MSE, MSLE, RRMSE, SSE, AMRE, MARE, SARE, MeAPE, MSRE, and RVE (see Table S1, available online). These metrics were selected based on the classification of Hauduc *et al.* (2015) as the combination of different metrics from different classes have been observed to be more effective than choosing metrics from one class only (Van Hoey 2016). In this way, each metric evaluates the proximity of two time series from a different perspective and the 12 metrics provide a full evaluation of the model performance against measured values. It is to be noted that the full informative potential of the 12 metrics is used in the dynamic simulation step. All metrics were chosen also based on their response range of values, and all indicate the best fit possible at 0. The metrics were selected based on their input requirements so that only values of observed and modelled results must be used as input. In this way, the response value of each metric can be rescaled based on its output from a minimum of 0 (best fit) to a maximum of 1 (worst fit).

Table 1 | Initial parameter selection showing extreme values of the domain used in literature

Parameter	Description	Minimum value	Maximum value
$K_{O\_A1Lysis}$ [g/m^3]	Saturation/inhibition coefficient for O_2 in lysis, AOB	0.2	1.6
$K_{O\_A2Lysis}$ [g/m^3]	Saturation/inhibition coefficient for O_2 in lysis, NOB	0.2	0.69
b_{A1} [1/d]	Rate constant for lysis of X_BA1	0.028	0.28
b_{A2} [1/d]	Rate constant for lysis of X_BA2	0.028	0.28
$n_{NOx\_A1\_d}$ [-]	Anoxic reduction factor for decay, AOB	0.006	0.72
K_{FA} [g/m^3]	Half-saturation index for Free Ammonia (FA)	0.001	0.005
K_{FNA} [g/m^3]	Half-saturation index for Free Nitrous Acid (FNA)	5.00E-07	5.00E-06
K_{I10FA} [g/m^3]	FA inhibition coefficient, NO_2 oxidation by NOB	0.5	1
K_{I10FNA} [g/m^3]	FNA inhibition coefficient, NO_2 oxidation by NOB	0.036	0.1
K_{I9FA} [g/m^3]	FA inhibition coefficient, NH_4 oxidation by AOB	0.1	1
K_{I9FNA} [g/m^3]	FNA inhibition coefficient, NH_4 oxidation by AOB	0.001	0.1
K_{OA1} [g/m^3]	O_2 half-saturation index for AOB	0.4	0.6
K_{OA2} [g/m^3]	O_2 half-saturation index for NOB	1	1.2
Y_{A1} [gCOD/gN]	Yield for AOB	0.15	0.24
Y_{A2} [gCOD/gN]	Yield for NOB	0.06	0.24
K_{FA_AOBden} [g/m^3]	NH_4 half-saturation index for AOB denitrification	0.001	1
K_{FNA_AOBden} [g/m^3]	FNA half-saturation index for AOB denitrification	1.00E-06	0.002
K_{IO_AOBden} [g/m^3]	Inhibition coefficient for O_2 in AOB denitrification	0	10
K_{SNO_AOBden} [g/m^3]	NO saturation coefficient for AOB denitrification	0.1	3.91
K_{SO_AOBden} [g/m^3]	O_2 sat coefficient for AOB denitrification	0.13	12
$n1_{AOB}$ [-]	Growth factor for AOB in denitrification step 1	0.08	0.63
$n2_{AOB}$ [-]	Growth factor for AOB in denitrification step 2	0.08	0.63
K_{A1} [g/m^3]	S_A saturation coefficient for heterotrophs aerobic growth	4	20
K_{F1} [g/m^3]	S_F saturation coefficient for heterotrophs aerobic growth	4	20
K_{O1_BH} [g/m^3]	Saturation/inhibition coefficient for heterotroph growth	0.2	1

Finally, the different scenarios were ranked based on the 0 to 1 value of each metric separately. Subsequently, an overall ranking can be derived based on the score that each scenario has in each of the metrics. In this way, each metric is scalable within its own domain to a 0 to 1 domain, and attaching to each scenario a value from 0 to 1 allows the ranking of the scenarios according to the single metric (Figure 2, left). The value that each scenario collects from each metric can then be summed up with the rest of the scores obtained from the rest of the metrics to obtain a final overall score used for the final ranking of a given scenario (Figure 2, right). The scenarios performing the best for all metrics, i.e. scoring nearly 0 for each different metric, result in the lowest overall score. The best one third of all the scenarios was selected as the *good scenarios*.

For the steady state case, the average output of the last part of the 100 days simulation (about 100 data points), were compared against the measured value. Therefore, for the evaluation of the steady state simulation outputs, the metric evaluation is only based on the proximity of two single values, i.e. the modelled mean and the relative reference value for NH_4, DO and TSS.

For the case of dynamic simulations, the model outputs of NH_4, NO_3, N_2O, and DO were compared against measured values. In this case, the metric evaluation becomes more complex due to the different nature of the metrics involved. Each metric will return an estimation of the performance of the model output giving more emphasis to different aspects of a model fit. Hence, the necessity of using the proposed ranking strategy summarizing the different aspects of the evaluation of a fit.

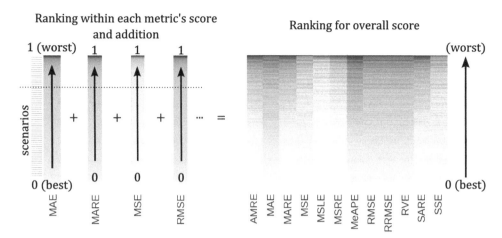

Figure 2 | Schematic representation of the scenario ranking method used in Step II and Step III. The initial ranking according to the single metric (left) allows to sum the scores of all metrics for each scenario and have a final score that is used for the overall ranking (right).

RESULTS AND DISCUSSION

Parameter ranking (Step I)

For the GSA with LH-OAT approach, different perturbation factors were investigated (De Pauw & Vanrolleghem 2006) resulting in a good performance of a 10^{-3} perturbation factor for all the parameters (Figures S1 and S2, available with the online version of this paper).

For the case of the TIS layout, the GSA results provided a ranking of the most influential parameters for N_2O, O_2, NO_3, NH_4, TSS, X_{BA1}, X_{BA2} and X_H. The influential parameters were the same among the different variables tested, showing very comparable overall importance with a few variations in the ranking (Table S2, available online).

The GSA exercise was repeated in the same way for the case of the CM layout. Interestingly, the relevant parameters were very similar to the case of the TIS showing very few variations and negligible differences from the previous ranking exercise.

Initially, 10 parameters (i.e. b_{A1}, b_{A2}, K_{OA1}, K_{FA}, K_{FNA}, K_{F1}, K_{OA2}, K_{O1_BH}, Y_{A1}, $n_{NOx\_A1\_d}$) were selected according to the score and visual analysis of the tornado plots (Rose 2013). Given the presence of Y_{A1}, the proximity in the ranking of Y_{A2}, and the attention that this parameter received in literature (*inter alia:* Pocquet *et al.* 2015; Spérandio *et al.* 2016), it was chosen to include also Y_{A2}. In a similar way, $K_{O\_A1Lysis}$ and $K_{O\_A2Lysis}$ were included given their strong link with K_{OA1}, their importance in the literature, and their proximity to the cut-off threshold of the sensitivity results. Finally, given the importance of K_{I9FA} in the process of N_2O production, it was also included. This selection resulted in a total of 14 parameters to be passed to Steps II and III.

In general, it is interesting that decay parameters for autotrophs are the most influential, and that a significant quantity of half-saturation indexes (K-values) are present in the ranking. This highlights the importance of the correct definition of half-saturation indexes (Arnaldos *et al.* 2015).

Steady state simulations (Step II)

The aim of Step II was to define the best scenario (i.e. set of parameters) for initializing the model for dynamic simulation (Step III) and to verify that the domain chosen for the different parameters was still valid, i.e. not indicating a need for a modification of the domain.

The output of the simulations, compared against average typical concentrations of NH_4, DO and TSS, making use of the selected metrics. This is thus a point-to-point comparison to make a preliminary selection of parameters ranges. Model outputs were scored from 0 (best) to 1 (worst) using the 12 metrics described and ranked accordingly in order to isolate the best performing scenarios.

TIS

Distribution plots of the parameter values relative to the best performing scenarios were used as an indication for possible reduction or modification of the parameter ranges adopted for these simulations before moving to Step III.

The results reflected the general tendency of abating K_{FNA} to very low values (normally in the order of 10^{-6}) and confirms the reported difficulties in the calibration of this parameter (Spérandio *et al.* 2016). Similarly, K_{FA} showed a perceivable preference towards lower values in its range.

From the isolation of the best performing scenarios according to the analysis of the modelled TSS, NH_4 and DO confirms the necessity of shifting the parameter range towards bigger values for b_{A1}.

By merging the three groups of best performing scenarios (i.e. for NH_4, DO, and TSS) it was possible to obtain an overall group of best performing scenarios. This overall group was used as ultimate check to include the parameter domains isolated for NH_4, DO, and TSS, and to help in defining whether the information gathered from the individual cases still holds when considering multiple parameters simultaneously.

CM

The visual ranking of the scenarios for the steady state simulations with the CM layout resembles very closely the ranking observed for the TIS layout (SM). This means that the absolute variations of the model output for the different scenarios are similar for the two layouts for NH_4, DO and TSS.

From the NH_4, DO, and TSS rankings individually and overall (merging the three groups of best performing scenarios), a clear tendency of b_{A1} to show a higher frequency in the highest part of its domain was noticed (Figure S9, available online). This was a clear indication of the need for a redefined domain for b_{A1} before passing to Step III. Similarly, b_{A2}, K_{F1}, and K_{FNA} returned a clear preference of the highest frequency of their distribution plot for the lower part of their domain.

Redefinition of parameter domains

According to the results of Step II for the TIS and CM layouts, some of the parameters would benefit from a modification of their sampling domain before passing to Step III. For those parameters showing truncated distributions and high frequency of best performing values close to an edge of their domain, the modification was considered. This reduces the number of simulations that likely result in a less good prediction and are not very useful in the analysis anyway. The domains of b_{A1}, Y_{A1}, K_{FA}, and K_{F1} were modified and values are reported in the SM (Table S3, available online).

Given the known tendency reported in literature for removing K_{FNA} values close to zero in order to accomplish a model fit, and given that those values are recognized to be unrealistic, the domain of K_{FNA} was not modified. In addition to this, the modification of the domain of four other parameters could already have a positive effect on K_{FNA}.

Dynamic simulations (Step III)

The model was initialized with a steady state simulation of 100 days for performing the dynamic simulations. A best scenario for initialization was thus needed. Using the intersection of the three groups of best performing scenarios, i.e. the scenario considered the best at the same time for NH_4, DO, and TSS could be identified.

Step III was targeted at defining the best performing scenarios by analyzing the dynamic simulation outputs against measured data in specific locations of the bioreactor. The aim of this phase was to compare the capabilities of the TIS and CM layouts in defining a set of scenarios best resembling the full-scale measured data. In this view, the scenarios were ranked according to the 12 metrics and compared, as in Step II, in terms of their capability of providing a realistic parameter range of best performing values. Therefore, the ranking used the same method as for Step II, but using online measured data for the metric comparison (i.e. NH_4, DO, N_2O and NO_3).

It must be pointed out that in the case of N_2O it was not possible to use all 12 metrics due to the fact that some metrics use the value at the denominator of a fraction returning an infinite solution if a variable reaches zero. AMRE, MARE, MSLE, MSRE, and SSE were not considered for ranking the scenarios according to the N_2O output.

For the TIS layout, the ranking according to the measured NH_4 (Figure 3) showed an interesting behavior of the MSLE metric which, at first sight, seems to rank the scenarios inversely to the rest of the metrics. This is true for some of the worst performing scenarios for MSLE (darker color), which are not considered as bad by the rest of the metrics. The reason lies in the high sensitivity of the MSLE to small differences between modelled and measured values. In particular, when both measured and modelled variables are smaller than 1, the discrepancy is enhanced by the effect of the logarithm and the quadratic term in the MSLE. Thus, the importance of using multiple metrics is illustrated once more. Using multiple metrics of different nature allows to analyze and rank the scenarios from different points of view, but also to compensate for particular behavior of a single metric. Nonetheless, the visualization proposed in this work highlights the contribution of the single metric and relative potential limits.

Concerning the ranking according to DO, all metrics resulted in well-behaving simulations (small values for the 12 metrics) and overall agreeing in a common final ranking.

Although the ranking according to N_2O was forced to have fewer metrics (see above), the seven metrics retained

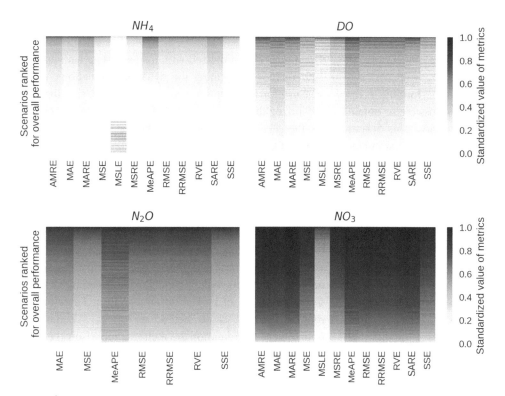

Figure 3 | Ranking of the scenarios (rows) according to the metrics (columns) from the best performing (bottom) to the worst (top). Each metric is colored according to its relative ranking from 0 to 1. Results of the TIS layout.

were still coming from different categories, thus ensuring a ranking according to different approaches. All metrics appear to rank similarly, although the fast transition towards the darkest colors suggests the presence of only a few scenarios performing significantly better than the rest.

In a similar picture, the ranking according to NO_3 can be observed, where, for most of the metrics, a fast transition to darker colors indicates a fast deviation of the modelled results away from the measured dataset.

Figure 4 shows the ranking for the scenarios of the CM layout. Small differences can be observed among the metrics for the ranking according to NH_4 in which MSLE seems to behave slightly different from the rest of the metrics, although generally agreeing with the rest of the metrics for the best performing scenarios (lighter colors).

For the case of DO there is faster transition to the darker tones of the ranking for all metrics, indicating probably that only a few scenarios are providing an output close to the measured dataset while the rest is quickly deviating away from it.

Differently from the case of DO, the case of N_2O and NO_3 present a very gradual shift away from the objective function making all metrics generally provide the same

ranking (note again that for N_2O fewer metrics could be considered).

Comparison between TIS and CM

The overall distributions of the parameter values for the best performing scenarios are derived from the ranking for NH_4, DO, N_2O, and NO_3, and selecting the best third of all the scenarios. These distributions are reported to make an overall comparison of the performances of both model layouts in defining ranges of parameter values that are best performing.

For the case of Y_{A1} (Figure 5), the CM configuration (right) returned a clearly defined range of acceptable parameter values as compared to the case of the TIS layout. The Y_{A1} histogram of the CM appears to recall a defined distribution curve which encounters a maximum frequency around the value of 0.1 g COD/g N. The TIS model (Figure 5, left) identifies the best performing scenarios in the lowest range of Y_{A1}, which are less realistic values as compared to the case of the CM.

K_{FNA} (Figure 6), is a parameter that is known to be difficult to calibrate, often leading to values very close to zero to force a fit (Spérandio *et al.* 2016). The CM results (Figure 6,

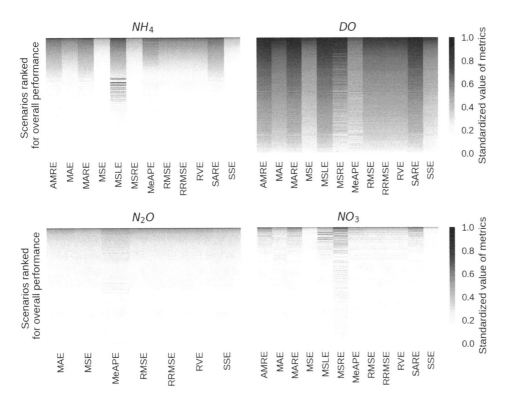

Figure 4 | Ranking of the scenarios (rows) according to the metrics (columns) from the best performing (bottom) to the worst (top). Each metric is colored according to its relative ranking from 0 to 1. Results of the CM layout.

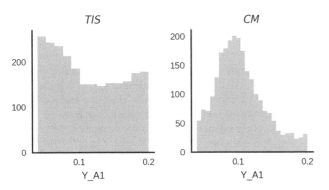

Figure 5 | Distributions of overall best performing scenarios for the case of the parameter values of Y_{A1} in the dynamic simulations with the TIS (left) and CM (left) layouts. Counts of good performing scenarios on the Y-axis.

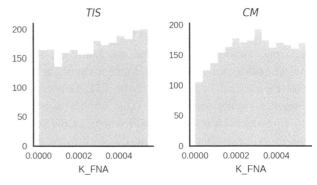

Figure 6 | Distributions of overall best performing scenarios for the case of the parameter values of K_{FNA} in the dynamic simulations with the TIS (left) and CM (left) layouts. Counts of good performing scenarios on the Y-axis.

right) again show a more pronounced shape of a distribution as compared to the TIS. This is an important result as it proposes more realistic values for this parameter and reverts the general tendency of moving this parameter down to 1E-6 g/m^3. On the other hand, the TIS layout does not show a definite distribution, having the same frequency almost everywhere. However, it must be pointed out that the frequency on the right edge of the distribution for the TIS is slightly increasing, suggesting the possibility of a need for a modification of the K_{FNA} domain.

In this view, it is interesting to consider that, despite literature studies generally reporting very low values of K_{FNA}, the TIS layout appears to revert this tendency showing this time a propensity for more realistic values. This might be because a rather consolidated TIS configuration is used for this plant, which has been calibrated and validated in detail in multiple past studies on this WRRF. Furthermore, it is interesting to point out how the CM model confirms the same tendency but with a more pronounced shape of the distribution. This is another confirmation that the higher hydrodynamic accuracy of the CM significantly increases the identifiability of some parameters.

CONCLUSIONS

In the present work, a ranking method and a visualization approach were proposed for selecting the best performing model parameter sets (scenarios) and providing a qualitative indication of the performance and contribution of each of 12 metrics used to express model fit. It is important to remember that this was not a calibration exercise and that model fits are deliberately not shown to emphasize the core messages.

The visual representation of the ranking of the different scenarios in both steady state and dynamic simulations returned interesting clues on the necessity of considering multiple metrics that assess different natures of model fit. Also, the performance of each metric was highlighted in its ranking and in the relative effect on the overall ranking of the scenarios, providing information on the contribution of the single metric and as a whole.

Plotting the parameter values obtained through the selection of the best performing scenarios, it was possible to directly compare the performance of the CM and TIS layouts in terms of parameter identification. The use of the CM increases the level of detail in the representation of local concentrations. The volume containing the sensors (and therefore providing local concentrations) is much better represented in the case of the CM. This improves calibration results significantly. This implies a relevant gain in model accuracy that redefined some of the key parameters to acceptable values and obtained more defined distributions.

The CM generally returned a more narrow parameter domain of good values compared to the TIS layout. This indicates that the more detailed description of local concentrations helps in defining a narrower domain of key parameters, which will, upon calibration, improve the model predictive power.

ACKNOWLEDGEMENTS

Peter A. Vanrolleghem, I. Nopens and L. Guo acknowledge the financial support obtained through the TECC project of the Québec Ministry of Economic Development, Innovation and Exports (MDEIE) and the joint research project funded by the Flemish Fund for Scientific Research (FWO-G.A051.10). Peter Vanrolleghem holds the Canada Research Chair on Water Quality Modeling.

REFERENCES

Amerlinck, Y. 2015 Model Refinements in View of Wastewater Treatment Plant Optimization: Improving the Balance in sub-Model Detail. Ghent University, Gent, Belgium.

Arnaldos, M., Amerlinck, Y., Rehman, U., Maere, T., Van Hoey, S., Naessens, W. & Nopens, I. 2015 From the affinity constant to the half-saturation index: understanding conventional modeling concepts in novel wastewater treatment processes. Water Research 70, 458–470.

Daelman, M. R. J., van Voorthuizen, E. M., van Dongen, U. G. J. M., Volcke, E. I. P. & van Loosdrecht, M. C. M. 2015 Seasonal and diurnal variability of N_2O emissions from a full-scale municipal wastewater treatment plant. Science of The Total Environment 536, 1–11.

De Keyser, W., Amerlinck, Y., Urchegui, G., Harding, T., Maere, T. & Nopens, I. 2014 Detailed dynamic pumping energy models for optimization and control of wastewater applications. Journal of Water and Climate Change 5, 299–314.

De Pauw, D. J. W. & Vanrolleghem, P. a. 2006 Practical aspects of sensitivity function approximation for dynamic models. Mathematical and Computer Modelling of Dynamical Systems 12, 395–414.

Gernaey, K. V. & Jørgensen, S. B. 2004 Benchmarking combined biological phosphorus and nitrogen removal wastewater treatment processes. Control Engineering Practice 12, 357–373.

Guo, L. 2014 Greenhouse gas Emissions From and Storm Impacts on Wastewater Treatment Plants: Process Modelling and Control. LAVAL University, Quebec City, QC, Canada.

Guo, L. S. & Vanrolleghem, P. a. 2014 Calibration and validation of an activated sludge model for greenhouse gases no. 1 (ASMG1): prediction of temperature-dependent N_2O emission dynamics. Bioprocess Biosyst. Eng. 37, 151–163.

Hauduc, H., Neumann, M. B., Muschalla, D., Gamerith, V., Gillot, S. & Vanrolleghem, P. A. 2015 Efficiency criteria for environmental model quality assessment: a review and its application to wastewater treatment. Environ. Model. Softw. 68, 196–204. https://doi.org/10.1016/j.envsoft.2015.02.004.

Hiatt, W. C. 2006 Activated Sludge Modeling for Elevated Nitrogen Conditions. Clemson University, Clemson, SC, USA.

Hiatt, W. C. & Grady, C. P. L. 2008 An updated process model for carbon oxidation, nitrification, and denitrification. Water Environment Research 80, 2145–2156.

IPCC 2013 Climate Change 2013: The Physical Science Basis. Contribution of Working Group I to the Fifth Assessment Report of the Intergovernmental Panel on Climate Change (T. F. Stocker, D. Qin, G.-K. Plattner, M. Tignor, S. K. Allen, J. Boschung, A. Nauels, Y. Xia, V. Bex & P. M. Midgley, eds). Cambridge University Press, Cambridge, UK and New York, NY, USA, 1535 pp.

Le Moullec, Y., Gentric, C., Potier, O. & Leclerc, J. P. 2010 Comparison of systemic, compartmental and CFD modelling approaches: application to the simulation of a biological reactor of wastewater treatment. *Chemical Engineering Science* **65**, 343–350.

Mampaey, K. E., Beuckels, B., Kampschreur, M. J., Kleerebezem, R., van Loosdrecht, M. C. M. & Volcke, E. I. P. 2013 Modelling nitrous and nitric oxide emissions by autotrophic ammonium oxidizing bacteria. *Environmental Technology* **34**, 1555–1566.

Ni, B. J. & Yuan, Z. 2015 Recent advances in mathematical modeling of nitrous oxides emissions from wastewater treatment processes. *Water Research* **87**, 336–346.

Ni, B. J., Ye, L., Law, Y., Byers, C. & Yuan, Z. 2013a Mathematical modeling of nitrous oxide (N$_2$O) emissions from full-scale wastewater treatment plants. *Environmental Science & Technology* **47**, 7795–7803.

Ni, B. J., Yuan, Z., Chandran, K., Vanrolleghem, P. a. & Murthy, S. 2013b Evaluating four mathematical models for nitrous oxide production by autotrophic ammonia-oxidizing bacteria. *Biotechnology and Bioengineering* **110**, 153–163.

Pocquet, M., Wu, Z., Queinnec, I. & Spérandio, M. 2015 A two pathway model for N$_2$O emissions by ammonium oxidizing bacteria supported by the NO/N$_2$O variation. *Water Research* **88**, 948–959.

Ravishankara, A. R., Daniel, J. S. & Portmann, R. W. 2009 Nitrous oxide (N$_2$O): the dominant ozone-depleting substance emitted in the 21st century. *Science* **326**, 123–125.

Rehman, U. 2016 *Next Generation Bioreactor Models for Wastewater Treatment Systems by Means of Detailed Combined Modelling of Mixing and Biokinetics*. Ghent University, Gent, Belgium.

Rehman, U., Maere, T., Vesvikar, M., Amerlinck, Y. & Nopens, I. 2014 Hydrodynamic-biokinetic model integration applied to a full-scale WWTP. In: *9th IWA World Water Congress and Exhibition*, Lisbon, Portugal.

Rehman, U., Vesvikar, M., Maere, T., Guo, L., Vanrolleghem, P. A. & Nopens, I. 2015 Effect of sensor location on controller performance in a wastewater treatment plant. *Water Science and Technology* **71**, 700.

Rehman, U., Audenaert, W., Amerlinck, Y., Maere, T., Arnaldos, M. & Nopens, I. 2017 How well-mixed is well mixed? hydrodynamic-biokinetic model integration in an aerated tank of a full-scale water resource recovery facility. *Water Science and Technology* **76**, 1950–1965.

Rose, K. H. 2013 A guide to the project management body of knowledge (PMBOK® guide), 5th edn. *Project Management Journal*.

Spérandio, M., Pocquet, M., Guo, L., Ni, B. J., Vanrolleghem, P. A. & Yuan, Z. 2016 Evaluation of different nitrous oxide production models with four continuous long-term wastewater treatment process data series. *Bioprocess Biosyst. Eng.* **39**, 493–510.

van Griensven, A., Meixner, T., Grunwald, S., Bishop, T., Diluzio, M. & Srinivasan, R. 2006 A global sensitivity analysis tool for the parameters of multi-variable catchment models. *Journal of Hydrology* **324**, 10–23.

Van Hoey, S. 2016 *Development and Application of A Framework for Model Structure Evaluation in Environmental Modelling*. Ghent University, Gent, Belgium.

Van Hulle, S. W. H., Callens, J., Mampaey, K. E., van Loosdrecht, M. C. M. & Volcke, E. I. P. 2012 N$_2$O and NO emissions during autotrophic nitrogen removal in a granular sludge reactor – a simulation study. *Environmental Technology* **33**, 2281–2290.

First received 7 August 2018; accepted in revised form 7 January 2019. Available online 17 January 2019